SDGSAT-1卫星微光影像图集

郭华东　主编

可持续发展大数据国际研究中心

科学出版社

北　京

内 容 简 介

本图集以 SDGSAT-1 卫星微光成像仪获取的 10 m 分辨率的彩色夜间灯光图像为主要内容，分别收录了亚洲、非洲、欧洲、美洲、大洋洲共计 147 个城市的微光影像及其文字介绍，展示各城市夜晚的繁华与活力，反映城市人居格局与经济发展状况，折射出不同地域、不同风俗、不同文化背景下各城市的民族风情和底蕴。

本图集既是一本具有科学价值和实用价值的微光遥感影像图集，也是一本具有艺术魅力和审美趣味的夜间灯光图集。通过 SDGSAT-1 卫星的眼睛，为读者提供一个独特的视角，观察世界。

本书可供遥感卫星技术和可持续发展科学研究领域的相关学者参考，也可作为地球科学及社会科学爱好者的兴趣读物阅读、收藏。

审图号：GS京（2023）1667号

图书在版编目（CIP）数据

SDGSAT-1 卫星微光影像图集 / 郭华东主编 . —北京：科学出版社，2023.9
ISBN 978-7-03-076208-5

Ⅰ. ①S… Ⅱ. ①郭… Ⅲ. ①遥感卫星—卫星图像—图集 Ⅳ. ① TP75-64

中国国家版本馆CIP数据核字（2023）第151503号

责任编辑：朱　丽　董　墨 / 责任印制：肖　兴
书籍设计：北京美光设计制版有限公司

科 学 出 版 社 出版

北京东黄城根北街16号
邮政编码：100717
http://www.sciencep.com

北京中科印刷有限公司 印刷
科学出版社发行　各地新华书店经销

*

2023年9月第 一 版　开本：889×1194 1/16
2023年9月第一次印刷　印张：18 1/2
字数：400 000

定价：518.00元

（如有印装质量问题，我社负责调换）

图集编辑委员会

主　　编：郭华东

副 主 编：窦长勇　　刘建波

编　　委：（按姓氏笔画排序）

前言

　　可持续发展是人类社会的永恒主题。自 1987 年可持续发展理念首次被提出，到 2000 年联合国千年发展目标，再到 2015 年联合国《2030 年可持续发展议程》（以下简称：2030 年议程），我们见证了人类在追求可持续发展道路上的坚定决心与不懈探索。然而，当今世界百年变局加速演进，全球经济复苏缓慢，极端气候和自然灾害频发，2030 年可持续发展目标的实现面临巨大的挑战。

　　以大数据、人工智能和空间技术等为代表的前沿科技正在重塑我们的生活，其中空间观测技术以其独特的优势，逐渐成为理解地球的新钥匙和知识挖掘的新手段，为服务全球可持续发展目标实现提供了全新视角。2018 年伊始，中国科学院设立了"地球大数据科学工程"A 类战略性先导科技专项（以下简称：地球大数据专项），专项设置了"可持续发展科学卫星 1 号（SDGSAT-1）"项目，专门服务可持续发展目标监测与评估的数据和技术需求。

　　SDGSAT-1 是全球首颗可持续发展目标观测科学卫星，于 2021 年 11 月 5 日成功发射。该卫星搭载了高性能微光、热红外和多谱段成像仪。通过三个载荷昼夜协同观测，实现对"人类活动痕迹的精细刻画"，揭示与人类活动和自然环境相关的可持续发展指标间的关联和耦合，以及人类活动引起的环境变化和演变规律，为表征人与自然交互作用的可持续发展目标研究提供数据支撑。这也标志着继我国相续推出气象卫星、资源卫星、环境卫星、海洋卫星和高分卫星等系列卫星之后，又一新的可持续发展系列卫星的问世。

　　城市作为人类社会活动的主要空间载体，其健康、和谐、可持续的发展，对于全人类走上可持续且具有韧性的发展道路和实现人与自然和谐相处具有至关重要的作用。SDGSAT-1 的技术特点可为城市可持续发展研究提供强有力的支撑：微光成像仪具备全球首创的同时获取 10 m 全色与 40 m 彩色微光的能力，热红外成像仪可通过三波段劈窗设计，以 30 m 分辨率识别地表 0.2℃的温度变化，多谱段成像仪可通过 1 个红边与 2 个深蓝波段的设计，分别监测植被生长状态和水质情况。同时，三个载荷的成像幅宽均为 300 km，保证了全球数据的获取能力。2022 年 9 月，卫星业主单位可持续发展大数据国际研究中心发起了"SDGSAT-1 开放科学计划"，同时中国政府宣布其数据向全球开放。现已有 20 余万景影像数据面向全球开放共享，已为来自 72 个国家和地区的用户和科研人员提供了数据支撑，助力各国可持续发展目标监测和决策支持。

作为全球首部城市夜间灯光影像图集，本书展示了 SDGSAT-1 微光成像仪获取的全球城市夜间彩色灯光影像，以全新视角生动细致地描绘了全球不同地区各城市的夜晚繁华与活力，折射出不同地域、不同风俗、不同文化背景下各城市的民族风情和底蕴。同时，反映出城市人居格局与经济发展状况，为研究可持续城市发展提供了宝贵的数据支持，展现出空间观测技术服务 2030 年议程实施的巨大潜力。

图集由六章组成，收录了全球包括亚洲、非洲、欧洲、美洲和大洋洲在内的 147 个城市彩色夜间灯光影像。第一章展示了中国京津冀、长三角、成渝和粤港澳大湾区 4 个经济圈 35 个主要城市的微光影像；第二章选择了 37 个亚洲首都城市的微光影像；第三章为 20 个非洲国家首都的微光影像；第四章展示了欧洲 37 个国家首都的夜间微光影像；第五章为南北美洲的 13 个代表性城市的微光影像；第六章展示了 5 个大洋洲城市的微光影像。这些微光影像为大家提供了一个了解全球不同地区城市夜间景象的机会，展示了城市的多样性、发展水平和文化特点。将帮助人们更好地了解不同城市的发展和规划，从而为实现城市的可持续发展目标做出贡献。

本图集是集体劳动的结晶。值全书付梓之际，笔者衷心感谢地球大数据专项领导小组组长白春礼、侯建国院长和副组长张亚平、张涛副院长，SDGSAT-1 卫星工程总指挥阴和俊和相里斌副院长、常务副总指挥于英杰局长、卫星工程总设计师樊士伟研究员，地球大数据专项各参研单位及专项项目一研究集体，SDGSAT-1 卫星工程研制队伍与运行团队，并特别感谢窦长勇、丁海峰、陈甫、尚二萍和龙腾飞博士等所做大量工作，感谢为本书做出贡献的所有人员。

受有关条件限制，书中影像质量存在差异，另由于时间仓促，书中难免存在疏漏与不妥之处，敬请同行专家和读者不吝指正。

中国科学院院士
SDGSAT-1 卫星首席科学家
2023 年 8 月 15 日于北京

Preface

    Sustainable development is the eternal pursuit of human society. Since the concept of sustainable development was first proposed in 1987, to the United Nations Millennium Development Goals (MDGs) in 2000, and then to the *United Nations 2030 Agenda for Sustainable Development* (2030 Agenda) in 2015, we have witnessed humankind's firm determination and tireless exploration on the path of pursuing sustainable development. However, the world is now facing great changes that have not been seen in a century with setbacks in the recovery of the global economy, escalating extreme climate, and frequent natural disasters, making it difficult for the world to make progress in the achievement of the 2030 Sustainable Development Goals (SDGs).

    Cutting-edge digital technologies, represented by big data, artificial intelligence and space technology, are reshaping our lives. Among them, Earth observation technology, with its unique advantages, has gradually become a new key to understanding the Earth system and a new means of knowledge mining, providing a new perspective to serve the implementation of the 2030 Agenda. At the beginning of 2018, the Chinese Academy of Sciences (CAS) set up the Strategic Pilot Science and Technology Project "Big Earth Data Science Engineering Program (CASEarth)", in which the project "Sustainable Development Science Satellite 1 (SDGSAT-1)" is tailored to serve the data and technology needs for monitoring and assessing the SDGs.

    SDGSAT-1, the world's first scientific satellite dedicated to serving the 2030 Agenda, was successfully launched on 5 November 2021. SDGSAT-1 is designed to carry three payloads including advanced Glimmer Imager, Thermal Infrared Spectrometer and Multi-spectral Imager. Through the coordinated day and night operations of the three payloads, the aim is to achieve a fine portrayal of the "traces of human activities" and provide data support for the study of SDGs characterizing the interaction between human beings and the natural environment. The SDGSAT-1 also marks the launch of a new series of satellites for sustainable development, following China's successive launch of meteorological satellites, resources satellites, environmental Satellites, marine satellite and high-resolution satellites.

As the main venue of human social activities, the healthy, harmonious and sustainable development of cities plays a crucial role in enabling all humankind to embark on a sustainable and resilient path of development and to achieve harmony between human beings and nature. The technical features of SDGSAT-1 can provide strong support for urban sustainable development research: the Glimmer Imager is the world's first spaceborne system that can acquire 10 m panchromatic and 40 m muti-band glimmer at the same time, the Thermal Infrared Spectrometer can identify the temperature change of 0.2 ℃ on the Earth surface with 30 m resolution through the designed the three-band split-window, and the Multispectral Imager monitors the growth status of the vegetation and the water quality through the design of the one red-edge band and the two deep-blue bands, respectively. The data from all three payloads have a swath of 300 km, which ensures the effectiveness of global data acquisition. In September 2022, the International Research Center of Big Data for Sustainable Development Goals (CBAS) , the operator of SDGSAT-1, launched the "SDGSAT-1 Open Science Program", and to date, more than 200,000 images acquired by SDGSAT-1 satellite have been shared globally free-of-charge to support research on sustainable development in various countries, and have provided data for researchers from 72 countries and regions to facilitate SDGs monitoring and decision making.

As the world's first atlas of nighttime city lights images, this book shows the multi-coloured light images of global cities at night acquired by SDGSAT-1 glimmer imager, which vividly depicts the prosperity of cities in different regions of the world at night from a brand-new perspective. These images show the national customs and heritage of cities from different geographical regions and cultural backgrounds. At the same time, they reflect the pattern of urban neighborhoods and economic development, providing valuable data support for the study of sustainable urban development and demonstrating the great potential of earth observation technology to serve the implementation of the 2030 Agenda.

The atlas consists of six chapters featuring multi-coloured nighttime light images of 147 cities around the world, including Asia, Africa, Europe, the Americas and Oceania. In the first chapter, the glimmer images of 35 major cities in China are shown; Chapter 2 selects 37 glimmer images of Asian capital cities; Chapter 3 presents glimmer images of 20 African capitals; Chapter 4 shows nighttime glimmer images of 37 European capitals; Chapter 5 is a collection of images from the Americas, with a selection of glimmer images from thirteen representative cities and Chapter 6 shows 5 cities in Oceania.

The atlas is a production of the team works. As the collection goes to press, the author would like to express his heartfelt thanks to CAS president Bai Chunli and Hou Jianguo, the head of leaders group of the CASEarth Program; CAS vice president Zhang Yaping and Zhang Tao, the deputy head of leaders group of the CASEarth Program; CAS vice president Yin Hejun and XiangLi Bin, the Chief Director of SDGSAT-1 satellite project; Yu Yingjie, the Deputy Chief Director of SDGSAT-1 satellite project; Professor Fan Shiwei, the Chief Designer of the SDGSAT-1 satellite project; The research, engineering and operation teams of the SDGSAT-1 satellite project; Especially, Dr. Dou Changyong, Dr. Ding Haifeng, Dr. Chen Fu, Dr. Shang Erping, Dr. Long Tengfei, etc. The author thanks all contributors to this atlas.

Due to inevitable adverse conditions, there are differences in the quality of the images in the book, and in addition, due to the haste of time, there are predictably some inconsistencies and inaccuracies in the book, so the author would like to invite peer experts and readers to suggest corrections.

Guo Huadong

CAS Academician
SDGSAT-1 Chief Scientist
Beijing, 15 August 2023

SDGSAT-1 卫星介绍

《联合国 2030 年可持续发展议程》（以下简称：2030 年议程）的实施迫切需要数据和方法的支撑。空间对地观测作为高效的数据获取手段和研究方法，能够为 2030 年议程做出重要贡献。为此，研制、运行系列可持续发展科学卫星成为可持续发展大数据国际研究中心的一项重要使命。

可持续发展科学卫星 1 号（SDGSAT-1）是全球首颗专门服务 2030 年议程的科学卫星，也是中国科学院研制并发射的首颗地球科学卫星。该卫星由中国科学院"地球大数据科学工程"A 类战略性先导科技专项（地球大数据专项）研制，是可持续发展大数据国际研究中心的首发星。

针对全球可持续发展目标（SDGs）监测、评估和科学研究的需求，SDG-SAT-1 卫星通过多载荷全天时协同观测，旨在实现"人类活动痕迹"的精细刻画，服务全球 SDGs 的实现，为表征人与自然交互作用的指标研究提供支撑。

SDGSAT-1 卫星通过探测人类活动与地球表层环境交互影响的地物参量，实现综合探测数据向 SDGs 应用信息的转化，研究跟人类活动和自然环境相关指标间的关联和耦合。充分利用 SDGSAT-1 卫星对地表进行宏观、动态、大范围、多载荷昼夜协同探测的优势，可以实现人居格局（SDG2、SDG6）、城市化水平（SDG11）、能源消耗（SDG13）、近海生态（SDG14、SDG15）等以人类活动为主引起的环境变化和演变规律研究，探索夜间灯光或月光等微光条件下地表环境要素探测的新方法与新途径，服务 SDGs 相关领域的研究。

SDGSAT-1 卫星为太阳同步轨道设计，搭载了高分辨率宽幅热红外、微光及多谱段成像仪三种载荷，轨道高度为 505 km，倾角为 97.5°，空间分辨率分别为 30 m、10 m/40 m 和 10 m，幅宽均为 300 km，重访周期约 11 天。设计有"热红外＋多谱段"、"热红外＋微光"以及单载荷观测等普查观测模式，可实现全天时、多载荷协同探测；同时，SDGSAT-1 拥有月球定标、黑体变温定标、LED 灯定标、一字飞行定标等星上和场地定标模式，保证了精确定量探测的需求。其中，热红外成像仪具有高分辨率宽幅观测能力，尤其在降低载荷功耗的前提下，空间分辨率比国际同类卫星提升 3.3 倍，是中国首次在热红外谱段采用全光路低温光学系统设计，可在大动态范围下实现优于 80 mK 的噪声等效温差。该卫星可精细探测城市热能分布，为测算全球能源消耗提供基础数据，服务清洁能源、气候行动等可持续发展目标。

微光和多谱段成像仪同样具备 300 km 的大幅宽，可实现地球昼夜全天时切换成像。其中微光成像仪是国际首个同时具备全色和彩色高分辨率微光探测载荷，可以 10 m（全色）和 40 m（彩色）空间分辨率探测夜间地表灯光的颜色、强度和空间分布，进而提供判识全球经济发展水平及区域发展差异的信息，服务可持续城市与社区、工业创新和基础设施等可持续发展目标。特别值得一提的是，SDGSAT-1 卫星过境时间为当地 21：00 至 22：00 之间，这一时段恰好是夜间人类活动的高峰期。与 Suomi NPP 卫星（其拍摄时间主要集中在当地时间凌晨 1:00 左右）相比，SDGSAT-1 获取的数据能更真实地反映人类活动的信息。多谱段成像仪拥有 7 个波段，具有高信噪比的特点，空间分辨率为 10 m，其中设计有 2 个深蓝波段和 1 个红边波段特别有利于水体质量和植被生长状态的监测，可服务清洁饮水和卫生设施、陆地生物等可持续发展目标。

SDGSAT-1 卫星具有以下特点：

（1）卫星平台：高载荷平台比，高灵敏度一体化设计与多模态高精度操控，高速数据技术及自主智能任务规划和标定模式。

（2）热红外成像仪：大幅宽（300 km）、高分辨率（30 m），幅宽／分辨率综合指标国际领先；高动态范围（220～340 K），高探测灵敏度（0.2K@300 K）。

（3）微光／多谱段成像仪：采用多模共光路成像设计，地球昼夜全天时切换成像；微光载荷探测动态范围 60 dB；多谱段载荷 7 个波段信噪比大于 130。

SDGSAT-1 卫星主要技术指标

类别	指标项	具体指标
轨道	轨道类型	太阳同步轨道
	轨道高度	505 km
	轨道倾角	97.5°
热红外成像仪	幅宽	300 km
	探测谱段	$8 \sim 10.5$ μm $10.3 \sim 11.3$ μm $11.5 \sim 12.5$ μm
	空间分辨率	30 m
微光/多谱段成像仪	幅宽	300 km
	微光探测谱段	P：$444 \sim 910$ nm B：$424 \sim 526$ nm G：$506 \sim 612$ nm R：$600 \sim 894$ nm
	微光空间分辨率	全色 10 m，彩色 40 m
	多谱段探测谱段	B1：$374 \sim 427$ nm B2：$410 \sim 467$ nm B3：$457 \sim 529$ nm B4：$510 \sim 597$ nm B5：$618 \sim 696$ nm B6：$744 \sim 813$ nm B7：$798 \sim 911$ nm
	多谱段空间分辨率	10 m

Introduction of SDGSAT-1

The implementation of the *United Nations 2030 Agenda for Sustainable Development* (2030 Agenda) urgently demands refined data and methodologies. Earth observation, an effective data collection and research method, holds significant potential for contributing to the 2030 Agenda. In view of this, the International Research Center of Big Data for Sustainable Development Goals (CBAS) has prioritized the development and operation of a series of scientific satellites dedicated to serving the global sustainable development.

The Sustainable Development Science Satellite 1 (SDGSAT-1) is a landmark achievement, being the first scientific satellite globally that is devoted to supporting the 2030 Agenda, and the premier Earth science satellite developed and launched by the Chinese Academy of Sciences (CAS). The development of SDGSAT-1 was facilitated by the "Big Earth Data Science Engineering" program of the CAS. SDGSAT-1 is the inaugural satellite in a series planned by CBAS.

SDGSAT-1 was designed for monitoring and evaluating the global sustainable development goals (SDGs), as well as facilitating related scientific research. Through multi-payload, round-the-clock collaborative observation, the satellite captures fine details of "human activity traces", aiding the realization of global SDGs and supporting research into the interplay between humanity and nature.

SDGSAT-1 effectively transforms comprehensive observation data into SDGs application information by detecting ground feature parameters concerning the interaction between human activities and the Earth's surface environment. It also examines the correlation and coupling with human activities and natural environment-related indicators. The satellite enables macroscopic, dynamic, wide-range, multi-payload, and 24-hour observation of the Earth's surface. This aids in studying environmental changes and evolution patterns brought about by human activities, such as human settlement patterns (SDG 2, SDG 6), urbanization levels (SDG 11), energy consumption (SDG 13), and near-shore ecology (SDG 14, SDG 15). It also explores innovative methods for detecting ground surface environmental elements under dim-light conditions, such as moonlight or nocturnal light, thereby assisting research related to the SDGs.

Operating in a sun-synchronous orbit, SDGSAT-1 is equipped with three types of payloads: high-resolution Thermal Infrared Spectrometer (TIS), Glimmer Imager (GI), and Multispectral Imager (MI). The satellite orbits at an altitude of 505 km, with an inclination angle of 97.5°, spatial resolutions of 30 m, 10/40 m and 10 m respectively, and a swath width of 300 km and revisit cycle of 11 days. It employs various observation modes, including a combination of TIS and MI, TIS and GI, as well as single-payload mode. These modes facilitate multi-payload coordinated observation throughout the day and night. Additionally, SDGSAT-1 utilizes on-orbit and vicarious calibration modes such as lunar calibration, blackbody calibration, LED calibration, and yaw maneuver calibration to ensure precise and quantitative applications.

The TIS has the ability of performing high-resolution (30 m) and wide-swath (300 km) observation, and its comprehensive performance has reached the international leading level. Remarkably, the spatial resolution is 3.3 times higher than that of similar international satellites, while also reducing payload power consumption. This is the first instance of China adopting an all-optical low-temperature system design in the thermal infrared spectrum, capable of distinguishing a temperature difference better than 80 mK in a large dynamic range. It can finely detect urban thermal energy distribution, provide fundamental data for calculating global energy consumption, and contribute to SDGs such as clean energy and climate action.

The GI and MI also have a large swath width of 300 km, enabling round-the-clock imaging of the earth through day and night. In particular, the GI is the world's first color high-resolution night-light detection payload, capable of detecting nighttime lights at a spatial resolution of 10m. In comparison, the resolution of glimmer images from other similar satellites is about 100~1000 m at present. It can provide night light intensity information to identify global economic development levels and regional development disparities, serving SDGs such as sustainable cities and communities and industrial innovation and infrastructure, etc. It's worth noting that SDGSAT-1's primary imaging time for GI is between 21:00 and 22:00 local time, coinciding with the peak of local human activities at night, making it particularly significant in terms of representativeness compared to other satellites like the Suomi NPP.

Meanwhile, the MI is characterized by a high signal-to-noise ratio, with a spatial resolution of 10 m. It is designed with seven bands, two of which are specifically designed in the deep blue spectrum, making it suitable for monitoring water color index, transparency, and suspended solids in various turbid water bodies; one of which is in red-edge band, ensuring the precise monitoring of growth status of vegetation on the ground. The SDGSAT-1 satellite encompasses the following performance features:

Satellite Platform: It has a high payload-to-platform ratio, a high-sensitivity integrated design, multi-mode high-precision control, high-speed data downloading technology, and autonomous intelligent task planning and calibration modes.

TIS: It has a wide swath width (300 km) and high resolution (30 m), leading the world in swath width/resolution. The imager has a wide dynamic range (220~340 K) and high detection sensitivity (0.2 K@300 K).

GI/MI: The imager adopts a multi-mode common-path imaging design and can alternatively switch imaging between the day and night. The dynamic range of glimmer payload detection is 60 dB. The signal-to-noise ratio of the 7 bands of the multispectral payload is greater than 130.

In summary, SDGSAT-1 embodies a significant step forward in sustainable development research. By providing a more nuanced understanding of the interaction between human activities and the natural environment, it offers invaluable insights and data to guide global efforts towards achieving the SDGs.

Technical Parameters of SDGSAT-1

Orbit/Sensors	Parameter	Specifications
Orbit	Type	Sun-synchronous
	Altitude	505 km
	Inclination	97.5°
Thermal Infrared Spectrometer	Swath Width	300 km
	Bands	8~10.5 μm 10.3~11.3 μm 11.5~12.5 μm
	Spatial Resolution	30 m
Glimmer/ Multispectral Imager	Swath Width	300 km
	Bands of GI	P: 444~910 nm B: 424~526 nm G: 506~612 nm R: 600~894 nm
	Spatial Resolution of GI	Panchromatic: 10 m, RGB: 40 m
	Bands of MI	B1: 374~427 nm B2: 410~467 nm B3: 457~529 nm B4: 510~597 nm B5: 618~696 nm B6: 744~813 nm B7: 798~911 nm
	Spatial Resolution of MI	10 m

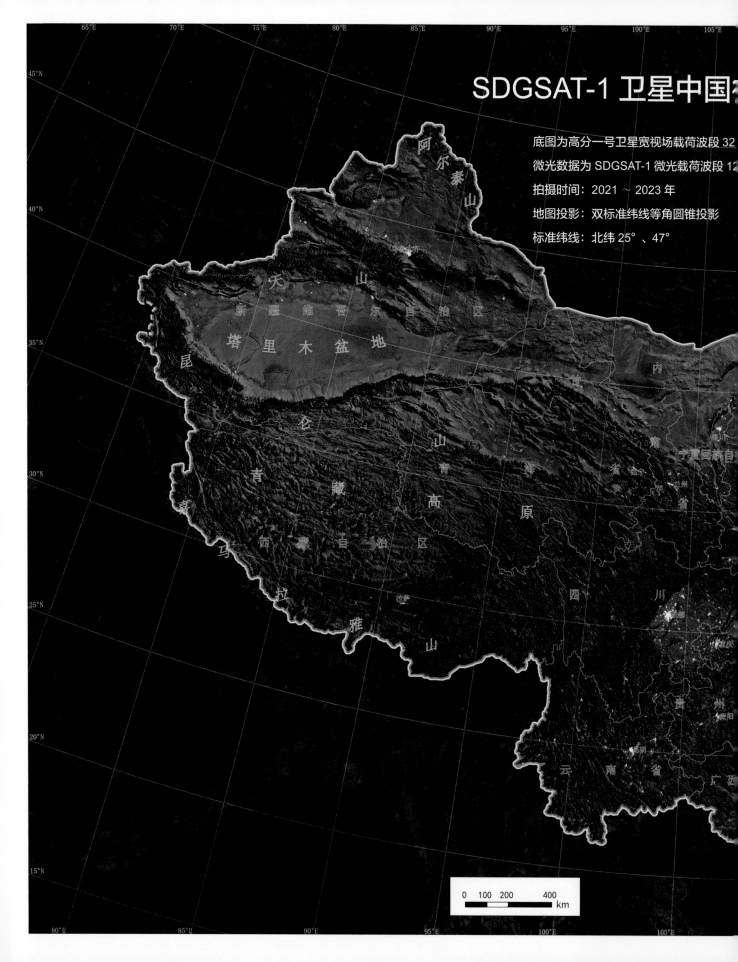

SDGSAT-1 卫星中国

底图为高分一号卫星宽视场载荷波段 32
微光数据为 SDGSAT-1 微光载荷波段 12
拍摄时间：2021 ～ 2023 年
地图投影：双标准纬线等角圆锥投影
标准纬线：北纬 25°、47°

0 100 200   400
                    km

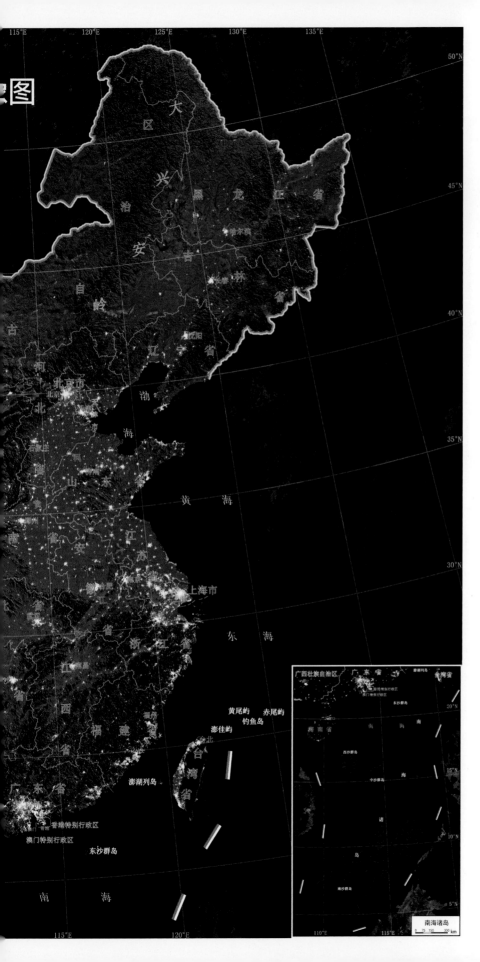

亚非欧大陆典型城

里斯本　　　布达佩斯　　　德黑兰

GSAT-1 卫星微光影像图

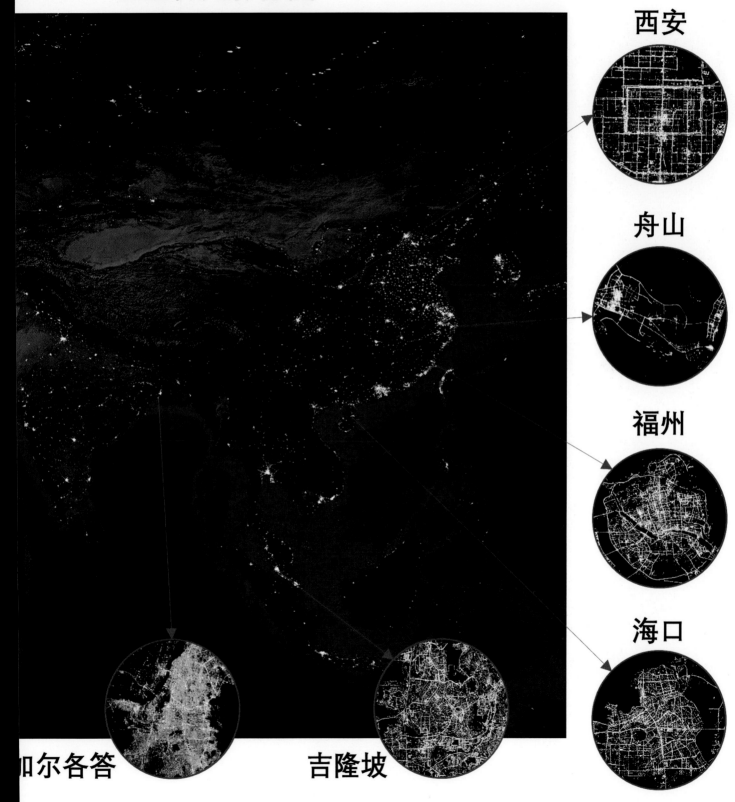

西安

舟山

福州

海口

加尔各答

吉隆坡

目录

中国

CHINA

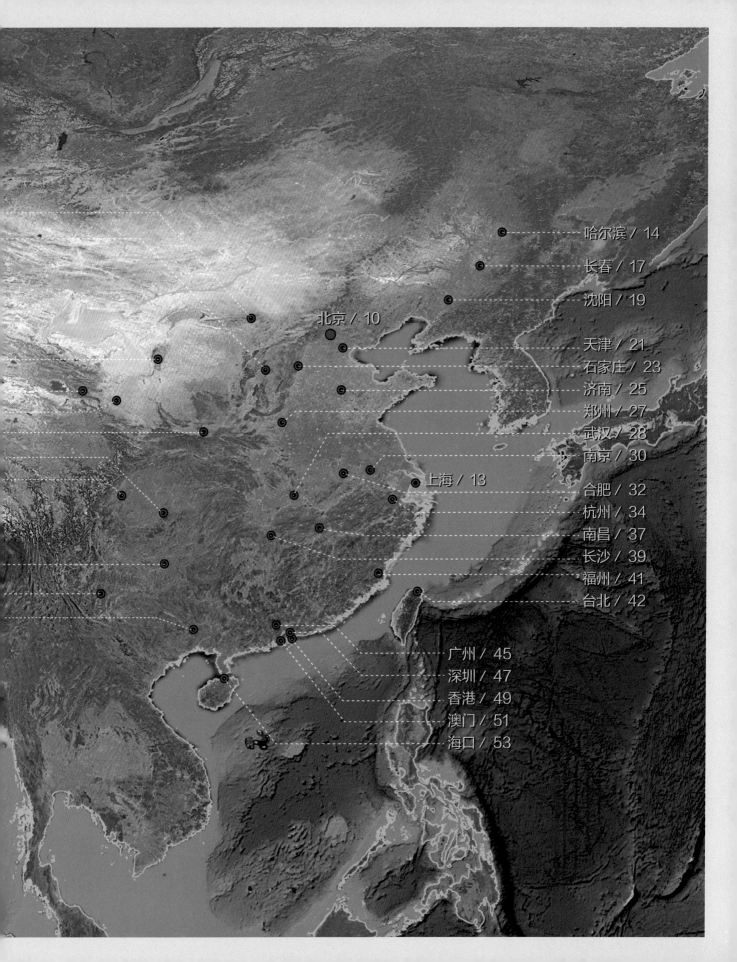

京津冀城市群
Beijing-Tianjin-Hebei urban agglomeration

京津冀城市群位于华北平原北部，是中国北方经济的重要核心区，总面积约为 218000 km^2，人口约为 1.1 亿。该地区拥有丰富的煤矿资源，是中国重要的能源和原材料供应基地，也是中国经济发展最具潜力的区域之一。2022 年，其 GDP 总量达到 10 万亿元，占全国 GDP 的 8.3%。并且，京津冀城市群正借助北京的研发创新优势、天津的制造业优势、河北的空间和人力资源优势，实现经济资源的优化配置和协调发展，打造具有全球竞争力的现代化城市群。京津冀城市群也是中国历史文化遗产和自然景观的集中地，如故宫、长城、山海关、白洋淀等，吸引众多中外游客前来欣赏。

The Beijing-Tianjin-Hebei urban agglomeration is the important core region of the economy in northern China, located in the northern part of the North China Plain, with a total area of about 218000 km^2 and a total population of about 110 million. The region has rich mineral resources such as coal and iron ore, and is an important base for energy and raw material supply in China. It is also one of the most potential regions for economic development in China. In 2022, its GDP reached 10 trillion CNY, accounting for about 8.3% of the national GDP. Moreover, the Beijing-Tianjin-Hebei urban agglomeration is leveraging the advantages of the capital's R&D and innovation, Tianjin's manufacturing industry, and Hebei's space and labor force to optimize and coordinate the allocation of economic resources and create a modern urban agglomeration with global competitiveness. The Beijing-Tianjin-Hebei urban agglomeration is also a concentration of historical and cultural heritage and natural landscapes in China, such as the Forbidden City, the Great Wall, Beidaihe, Shanhaiguan, Baiyangdian, etc., attracting many domestic and foreign tourists.

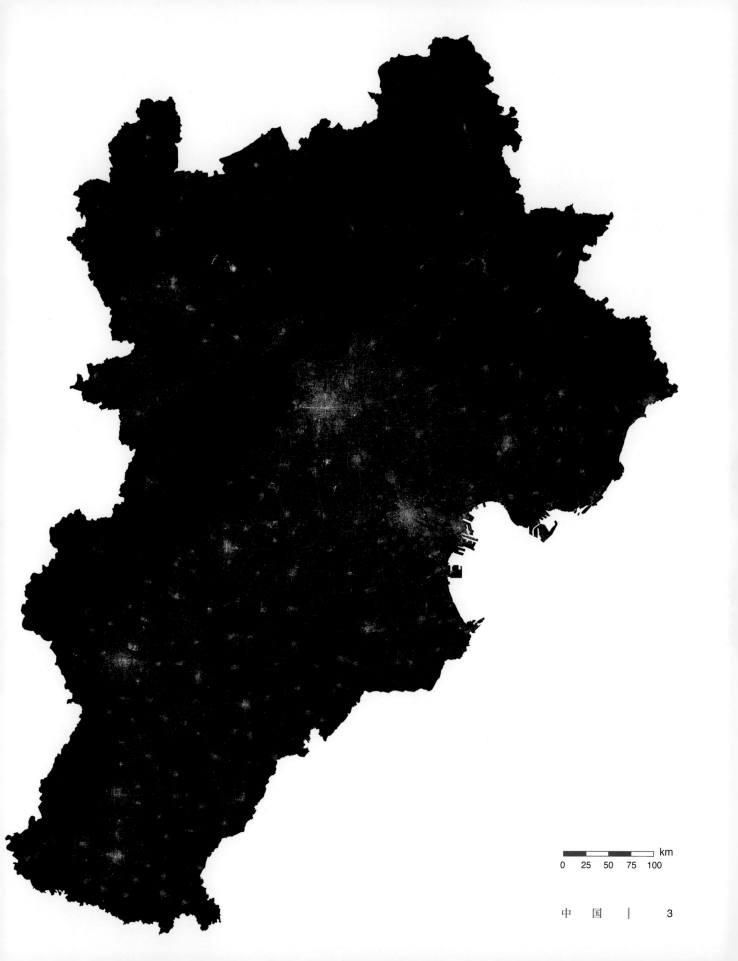

长江三角洲城市群
Yangtze River Delta urban agglomeration

　　长江三角洲城市群（简称长三角城市群）是中国东部沿海地区发展繁荣的经济区域，该城市群涉及一市（上海市）三省（江苏省、浙江省、安徽省），规划共 27 个市，覆盖面积约 211700 km²，人口约 2.37 亿。它是中国参与全球经济合作的主要区域，也是亚太地区重要的国际枢纽。2022 年，该区域的GDP 达到 24.5 万亿元，占全国的 20%。长三角城市群资源丰富、环境优良、交通发达，是中国科技创新与技术研发的领先区域，也是具有全球影响力的创新高地。长三角城市群拥有许多风景名胜和历史文化遗产，如外滩、西湖、乌镇、黄山等，是建设美丽中国的典范。长三角城市群是一个面向全球、辐射亚太、引领中国的世界级城市群，是中国现代化建设最好的地区之一。

　　The Yangtze River Delta urban agglomeration, nestled along the vibrant eastern coast of China, stands as a thriving economic region. Encompassing 27 prefecture-level cities across Shanghai, Jiangsu, Zhejiang, and Anhui provinces, this agglomeration spans an impressive area of about 211700 km² and hosts a population of approximately 237 million. It is an important area for China to participate in global economic cooperation, and an important international hub in the Asia-Pacific region. In 2022, its GDP reached 24.5 trillion CNY, accounting for 20% of the national total. The Yangtze River Delta urban agglomeration is a leading region for scientific and technological innovation and technology research and development in China, as well as a globally influential innovation highland. This region has many scenic spots and historical and cultural heritage, such as the Bund, West Lake, Wuzhen, Huangshan, etc., and is a pioneer of reform and opening up and a model of beautiful China construction. The Yangtze River Delta urban agglomeration is a world-class urban agglomeration that faces the world, radiates the Asia-Pacific region and leads China, and is one of the best regions for China's modernization construction.

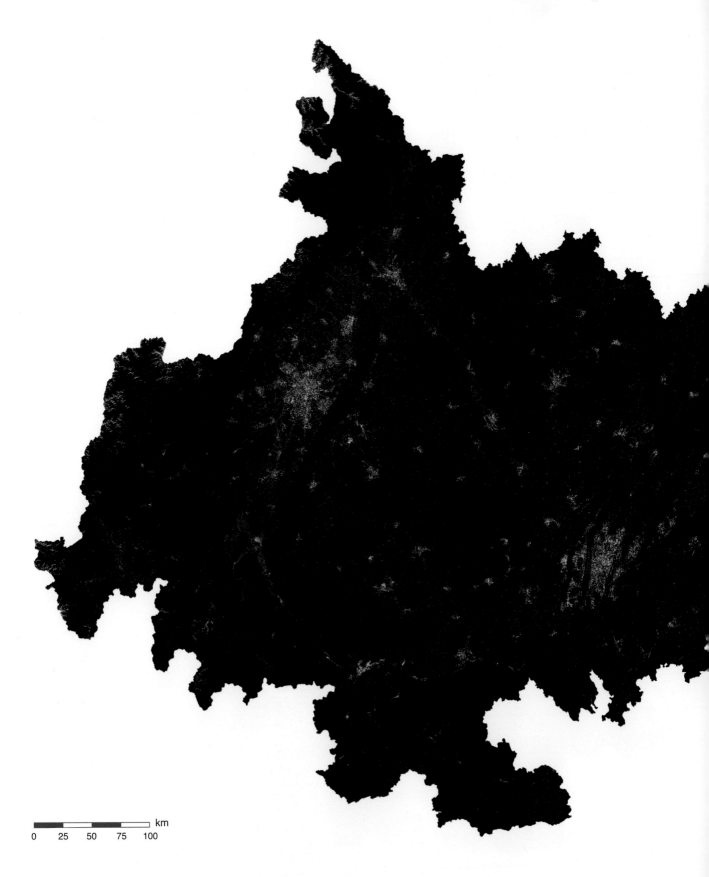

km

0   25   50   75   100

成渝城市群
Chengdu-Chongqing urban agglomeration

　　成渝城市群是中国西部人口最多、城镇化水平最高的区域之一，位于"一带一路"和长江经济带交会处，总面积约为 185000 km²。成渝城市群是中国内陆地区发展最迅速的地区之一，拥有丰富的人力、物力、财力资源，以及广阔的市场和良好的投资环境，在制造业、能源和交通等领域具有很强的优势，是中国的经济中心之一。该城市群以重庆和成都为核心，呈现双核心辐射型结构，此外，两核心相向发展，都具备发展现代制造业、服务业和创新科技的优势。周边其他城市具有不同程度的辐射带动作用，形成了明显的城市群和经济走廊。

The Chengdu-Chongqing urban agglomeration is one of the regions with the largest population and highest urbanization level in western China. It is located at the intersection of the "one belt, one road" and the Yangtze River Economic Belt, with a total area of about 185,000 km². The Chengdu-Chongqing urban agglomeration is one of the fastest-growing regions in inland China, with abundant human, material, and financial resources, as well as a broad market and a favorable investment environment. The region has strong advantages in manufacturing, energy, transportation, and other fields, making it one of the economic centers in mainland of China. The Chengdu-Chongqing urban agglomeration shows a dual-core radiation structure, in which Chongqing and Chengdu develop towards each other. Both cities have advantages in developing modern manufacturing, service, and innovative technology. Other surrounding cities also play a role in driving economic development to varying degrees, forming a distinct urban agglomeration and economic corridor.

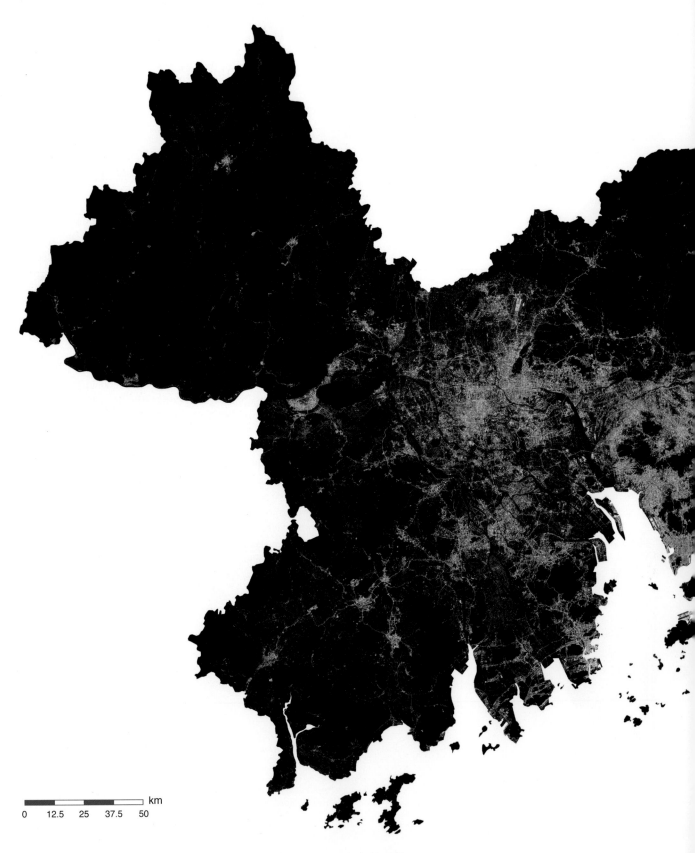

粤港澳大湾区城市群
Guangdong-Hong Kong-Macao Greater Bay Area urban agglomeration

　　粤港澳大湾区位于中国南部，涵盖广东省、香港特别行政区和澳门特别行政区，是一个以珠江三角洲为核心的经济合作区域。该湾区具有独特的区位优势，是继美国纽约湾区、美国旧金山湾区、日本北部湾区之后世界第四大湾区，也是连接中国和世界重要门户。粤港澳大湾区总面积约为 55900 km²，人口总量约为 0.87 亿。该湾区以珠江三角洲为核心，通过深化区域合作和互联互通，实现了空间上的紧密联系。如今，粤港澳大湾区覆盖 11 个城市的立体交通网络正越织越密，"1 小时生活圈"基本形成。这里汇集了广东省制造业基础、香港国际金融中心和澳门旅游娱乐业等丰富的经济资源，经济实力雄厚，创新资源丰富，是中国对外开放的重要窗口。

The Guangdong-Hong Kong-Macao Greater Bay Area is located in southern China, covering Guangdong Province, Hong Kong Special Administrative Region, and Macao Special Administrative Region. It is an economic cooperation region with the Pearl River Delta as its core. The bay area is the fourth largest bay area in the world and an important gateway connecting mainland of China and the world. Its total area is approximately 55900 km², and the total population is about 87 million. The bay area has achieved close spatial connections through deepening regional cooperation and interconnection, forming a "one-hour living circle". The Guangdong-Hong Kong-Macao Greater Bay Area gathers rich economic resources such as Guangdong Province's manufacturing base, Hong Kong's international financial center, and Macao's tourism and entertainment industry. It has strong economic strength, abundant innovation resources, and is an important window for China's opening up.

中国 北京
Beijing, China

　　北京是中国的首都，也是世界上人口最多的国家首都，2022 年，常住人口约 2184 万人。北京全市总面积 16410.54 km^2，与河北省、天津市相邻，是京津冀城市群和国家首都功能核心区的一部分。北京是一座全球性城市，是世界文化、外交、政治、金融、商业等领域的重要中心。作为中国四大古都之一，北京保留了众多的历史文化古迹。北京的紫禁城、天坛、颐和园、十三陵、周口店、长城和京杭大运河这 7 处世界文化遗产吸引了无数中外游客。北京还是世界上首个"双奥之城"，成功举办了 2008 年夏季奥运会以及 2022 年冬季奥运会，主体育场"鸟巢"是北京的标志性建筑之一。

Beijing is the capital of the People's Republic of China and the world's most populous national capital, with a total population of about 21.84 million (2022). Beijing covers a total area of 16410.54 km^2 and borders Hebei Province and Tianjin Municipality. It is part of the Beijing-Tianjin-Hebei urban agglomeration and the core area of the national capital function. Beijing is a global city and an important center of world culture, diplomacy, politics, finance, business and other fields. Beijing Daxing International Airport is Beijing's second international airport, with the world's largest single-structure airport terminal building and one of the busiest airports in the world in terms of passenger traffic. As one of the four ancient capitals of China, Beijing preserves many historical and cultural relics. The seven world cultural heritage sites in Beijing, such as the Forbidden City, Temple of Heaven, Summer Palace, Ming Tombs, Zhoukoudian, Great Wall and part of the Grand Canal, attract countless domestic and foreign tourists. Beijing is the first "double Olympic city" city in the world, successfully hosting the 2008 Summer Olympics and the 2022 Winter Olympics. The Olympic venue "Bird's Nest" has also become one of Beijing's iconic buildings.

中国 上海
Shanghai, China

    上海，中国四个直辖市之一，是国家中心城市和超大城市，也是中国的经济、金融、贸易、航运、科技和工业中心，位于长江入海口，南临杭州湾，北、西与江苏、浙江两省相接，隔东海与日本九州岛相望。上海下辖 16 个行政区，总面积 6340.5km²，是全球规模最大的都市之一。2022 年，上海市常住人口为 2475.89 万人。上海与江苏、浙江、安徽共同构成的长江三角洲城市群已成为国际六大世界级城市群之一，是中国大陆首个自贸区"中国（上海）自由贸易试验区"所在地。

    Shanghai, one of the four municipalities directly under the central government of China, is a national central city, a megacity, and an important center for economic, financial, trade and other fields in China. It is located at the mouth of the Yangtze River, bordering Jiangsu, Zhejiang and Anhui provinces, and facing Japan's Kyushu Island across the East China Sea. Shanghai has 16 municipal districts under its jurisdiction, with a total area of 6340.5 km². It is one of the largest metropolitan areas in the world in terms of scale and area. By the end of 2022, Shanghai's permanent resident population was 24.7589 million. The Yangtze River Delta urban agglomeration composed of Shanghai. Jiangsu, Zhejiang and Anhui have become one of the six world-class urban agglomerations in the world. Shanghai Port ranks first in the world in both cargo throughput and container throughput. At the same time, Shanghai is also the location of the first free trade zone in mainland of China, the "China (Shanghai) Pilot Free Trade Zone".

km
0 2 4 6 8

中国 哈尔滨
Harbin, China

哈尔滨是黑龙江省的省会，是中国东北部最大和人口最多的城市之一，2022年，人口约939.5万，面积为53076.4 km²。这座城市是中国东北地区重要工业城市，工业基础雄厚，在工程机械、石油化工食品加工等领域有着突出优势。哈尔滨文化独特历史厚重，有丰富的旅游景点，包括著名的太阳岛、圣索菲亚教堂和中央大街等，吸引着来自世界各地的游客。

Harbin is the capital of Heilongjiang Province and one of the largest and most populous cities in northeastern China, with a population of about 9.395 million (2022) and an area of 53076.4 km². The city is an important industrial city in the northeastern China, with a strong industrial base and outstanding advantages in engineering machinery, petrochemical, food processing and other fields. Harbin has a unique and rich culture and history, with abundant tourist attractions, including the famous Sun Island, St. Sophia Cathedral, Central Street and, etc., attracting tourists from all over the world.

km
0   1.25   2.5   3.75   5

中国 长春
Changchun, China

　　长春是吉林省的省会，东北亚经济圈中心城市，也是我国重要的工业基地。长春位于吉林省中部，是东北三省的中心城市之一，地处松辽平原，气候宜人，四季分明，夏季凉爽，冬季寒冷。长春是国家历史文化名城，具有丰富的近现代历史，拥有众多工业遗产和文化遗存；是东北地区重要的经济中心，拥有发达的制造业和服务业，如汽车制造、机械制造、医药制造等。长春的蓬勃发展也吸引了许多游客亲自前往长春，感受其历史文化和现代科技的独特魅力。2022 年，全市总面积 24592 km²，常住人口 908.72 万人。

　　Changchun is the capital of Jilin Province, a hub in the Northeast Asia economic circle, and an important industrial base in China. Located in the central part of Jilin Province, Changchun is one of the key cities in the three northeastern provinces. The city is situated on the Songliao Plain, with distinct seasons, cool summers and cold winters. As a national historical and cultural city, Changchun has a rich modern history, and is home to numerous industrial heritage sites and cultural relics. Changchun is an important economic center in the northeast region, with developed manufacturing and service industries, such as automobile manufacturing, machinery manufacturing, pharmaceutical manufacturing and so on. Changchun's economic development also attracts many tourists to visit, experience Changchun's historical culture and modern technology, and discover its unique charm. As of 2022, Changchun has a total area of 24592 km² and a resident population of 9.0872 million people.

km
0　　1.25　　2.5　　3.75　　5

中国 沈阳
Shenyang, China

　　沈阳是辽宁省的省会，位于辽宁省的中北部，是辽宁省人口最多的城市。沈阳的西部位于辽河流域的冲积平原，而东部则拥有丰富的森林资源。沈阳是中国重要的工业基地，是沈阳经济区的核心城市。该市的工业以重工业为主，特别是航空航天、机床、重型设备和国防产业，近年来也发展了软件、汽车和电子产品等行业。沈阳还是历史文化名城，清朝发祥地，素有"一朝发祥地，两代帝王都"之称，拥有沈阳故宫、清昭陵、福陵、张氏帅府和中国工业博物馆等众多文化遗产和旅游景点。沈阳具有丰富的资产和规模优势，独特的创新资源，以及雄厚的产业基础，其人才资源丰富，创新成果突出，未来的产业发展充满战略性和独特性。截至 2022 年底，全市总面积 12860 km²，常住人口 914.7 万。

Shenyang is the capital of Liaoning Province, located in the north-central part of Liaoning, and the most populous city in the province. The western part of Shenyang is characterized by the alluvial plains of the Liao River basin, while the east boasts rich forest resources. Shenyang is an important industrial base in China and the core city of the Shenyang Economic Zone. Shenyang's industry is dominated by heavy industry, especially aerospace, machine tools, heavy equipment and defense industries. In recent years, it has also developed software, automotive and electronic products industries. Shenyang is a historical and cultural city and the birthplace of the Qing Dynasty, known as "the place where a dynasty originated and two emperors reigned". Shenyang has many cultural relics and tourist attractions, such as the Shenyang Imperial Palace, Zhaoling Mausoleum, Fuling Mausoleum, Zhang's Mansion, China Industrial Museum and so on. Shenyang has a strong asset stock and scale advantage, with distinctive innovation resources solid industrial foundation, rich talent resources and innovation achievements. It has strategic and characteristic advantages for developing future industries. By the end of 2022, Shenyang has a total area of 12860 km² and a resident population of 9.147 million people.

km
0　1.25　2.5　3.75　5

中国 天津
Tianjin, China

　　天津是中国北方最大的沿海城市，位于渤海之滨，是中国四个直辖市之一。天津与河北省和北京市相邻，东濒渤海湾，是环渤海经济圈的一部分，也是京津冀城市群的重要组成部分。天津是一个"双核"城市，其主城区沿海河而建，海河通过大运河与黄河和长江相通；老城区东部的滨海新区是新兴的城市核心区域，位于渤海湾沿岸。海河穿城而过，高耸的天津之眼横跨其间。晚上漫步海河边，从摩天轮上俯瞰城市夜景的体验妙不可言。截至 2022 年底，全市总面积 11966.45 km²，常住人口 1363 万。

　　Tianjin is the largest coastal city in northern China, located on the shores of the Bohai Sea and is one of China's four municipalities directly under the central government. Tianjin borders Hebei Province and Beijing, and faces the Bohai Bay to the east. It is an important part of the Bohai Economic Circle, and also a member of the Beijing-Tianjin-Hebei metropolitan area. Tianjin is a dual-core city, with its main urban area built along the Haihe River, which connects to the Yellow River and the Yangtze River through the Grand Canal; Binhai New Area is another emerging urban core region, located east of the old city district and along the coast of the Bohai Bay. The Haihe River flows gently through the city, with the towering Tianjin Eye spanning over it. At nightfall, walking along the Haihe River and overlooking the night view from the Ferris wheel is an indescribable experience. As of the end of 2022, Tianjin has a total area of 11966.45 km² and a resident population of 13.63 million.

中国 石家庄
Shijiazhuang, China

　　石家庄是河北省的省会，位于太行山东部。该市地处河北省中南部，属环渤海经济圈，位于首都北京的西南方向，跨华北平原和太行山地两大地貌，是全国重要的商品集散地和北方重要的大商埠、全国性商贸会展中心城市之一。石家庄是华北地区的工业重镇，也是北京 - 天津 - 石家庄高科技产业带的一环。石家庄以医药业闻名，有"中国医药中心"之称，还有纺织、机械、化工、建材、轻工和电子等产业。石家庄农业发达，是河北省的棉花、梨、枣、核桃重要供应地。截至 2022 年底，全市总面积 14530 km²，常住人口 1122.35 万。

　　Shijiazhuang is the capital of Hebei Province, located on the eastern side of the Taihang Mountains. The city is located in the south-central part of Hebei Province and belongs to the Bohai Economic Circle. It lies to the southwest of China's capital, Beijing, spanning the North China Plain and the terrain of the Taihang Mountains. Shijiazhuang is an industrial hub in North China and also part of the Beijing-Tianjin-Shijiazhuang High-tech Industrial Belt. Shijiazhuang is famous for its pharmaceutical industry and is known as the "China Pharmaceutical Center". It also has industries such as textiles, machinery, chemicals, building materials, light industry and electronics. Shijiazhuang has a developed agriculture and is a major producer of cotton, pears dates and walnuts in Hebei Province. As of the end of 2022, Shijiazhuang has a total area of 14530 km² and a resident population of 11.2235 million.

中国 济南

Jinan, China

　　济南是山东省的省会，是中国 15 个副省级城市之一。作为环渤海的中心城市，济南是山东省新旧动能转换综合试验区核心区之一，也是山东省政治、经济、文化、金融中心。济南北邻京津冀经济圈，位于环渤海经济圈的西南部，是一座历经 2600 年沧桑的历史文化名城，也是一座集山、泉、河、湖与现代都市生活为一体的优秀旅游城市。济南以趵突泉、黑虎泉、五龙潭、珍珠泉四大名泉著称，拥有 10 个温泉群、72 个名泉、808 个天然温泉，素有"泉城"美誉。截至 2022 年底，全市总面积10244.45 km²，常住人口 941.5 万。

Jinan is the capital of Shandong Province and one of the 15 sub-provincial cities in China. As the central city of the Bohai Rim, Jinan is one of the core areas of the comprehensive pilot zone for new and old kinetic energy conversion in Shandong Province, and also the political, economic, cultural, and financial center of Shandong Province. Jinan is adjacent to the Beijing-Tianjin-Hebei economic circle in the north and in the southwest of the Bohai Economic Circle. Jinan is a historical and cultural city that has experienced 2600 years of vicissitudes, and also an excellent tourist city that integrates mountains, springs, rivers, lakes and modern urban life. Jinan is famous for its four famous springs Baotu Spring, Black Tiger Spring, Five Dragon Pool, and Pearl Spring. It has 10 hot spring groups, 72 famous springs, and 808 natural hot springs, earning it the nickname "Spring City". As of the end of 2022, Jinan has a total area of 10244.45 km² and has a resident population of 9.415 million.

km
0   1.25   2.5   3.75   5

中国 郑州

Zhengzhou, China

郑州是河南省的省会，是中国国家中心城市之一。作为河南省的最大城市，郑州地处黄河中下游的分界处，总体地形格局西南高、东北低，境内有 124 条河流，地跨黄河和淮河两大流域。郑州是河南省的政治、经济、科技和教育中心，郑州都市圈是中原经济区的核心区。郑州有着中国第一家期货交易所一郑州商品交易所 (ZCE)，以及中国首个空港经济区一郑州空港经济区。郑州有登封"天地之中"历史建筑群和中国大运河通济渠郑州段两处世界历史文化遗产。截至 2022 年底，全市总面积 7567 km²，常住人口 1282.8 万。

Zhengzhou is the capital of Henan Province and one of the national central cities in China. As the largest city in Henan Province, Zhengzhou is located at the dividing line of the middle and lower reaches of the Yellow River. with a general terrain of high in the southwest and low in the northeast. There are 124 rivers within it's territory, spanning both the Yellow River and the Huai River basins. Zhengzhou covers an area of 7567 km². Zhengzhou is the political, economic, scientific and educational center of Henan Province, and the Zhengzhou metropolitan area is the core area of the Central Plains Economic Zone. Zhengzhou has China's first futures exchange - Zhengzhou Commodity Exchange (ZCE), and China's first airport economic zone - Zhengzhou Airport Economic Zone. Zhengzhou has two World Cultural Heritage sites: historic monuments of Dengfeng in"The Centre of Heaven and Earth"and the Zhengzhou section of the Grand Canal of China. As of the end of 2022, Zhengzhou has a total area of 7567 km² and has a resident population of 12.828 million.

中国 武汉

Wuhan, China

　　武汉是湖北省的省会，是中国九个国家中心城市之一，这座历史悠久的城市位于江汉平原东部，地处长江及其最大支流汉江的交汇处，自古就是交通、经济、文化的重要枢纽。武汉的地理位置优越，拥有数十条铁路、公路，被称为"九省通衢"。武汉不仅是湖北省的政治和经济中心，还是中部地区的学术和研究中心。多所知名大学和研究机构驻扎于此，如华中科技大学和武汉大学。武汉的历史与文化遗产也相当丰富，从黄鹤楼到古琴台，都承载了这座城市深厚的历史底蕴。截至 2022 年底，全市总面积 8569.15 km²，常住人口 1373.90 万。

　　Wuhan is the capital of Hubei Province and one of the nine national central cities in China. This city, steeped in history, is located in the eastern part of the Jianghan Plain, at the confluence of the Yangtze River and its largest tributary, the Han River. Historically, Wuhan has been a crucial hub for transportation, economy, and culture. With its advantageous geographical location, Wuhan connects dozens of railways and highways, and is known as the "Thoroughfare of Nine Provinces". While Wuhan serves as the political and economic heartbeat of Hubei, it further distinguishes itself as the academic and research nucleus of the central region. Numerous renowned universities and research institutions, such as Huazhong University of Science and Technology and Wuhan University, are based here. The city's rich historical and cultural heritage is evident, with landmarks like the Yellow Crane Tower and the Guqin Terrace bearing testament to its profound legacy. As of the end of 2022, Wuhan has a total area of 8569.15 km² and has a resident population of 13.739 million.

中国 南京

Nanjing, China

　　南京是江苏省的省会，位于江苏省西南部、长江下游。作为中国东部地区重要的中心城市，南京是全国重要的科研教育基地和综合交通枢纽，是长三角辐射带动中西部地区发展的重要门户城市和东部沿海经济带与长江经济带战略交会的重要节点城市。南京是中国首批国家历史文化名城和全国重点风景旅游城市之一，从古老的秦淮河、明故宫到现代的新街口，南京的每一个角落都充满了故事。南京市平面呈"南北长东西窄"的格局，东西最大横距约 70 km，南北最大纵距约 150 km。截至 2022 年底，全市总面积 6587.02 km²，常住人口 949.11 万。

　　Nanjing is the capital of Jiangsu Province, located in the southwest of Jiangsu Province and the lower reaches of the Yangtze River. As an important central city in eastern China, Nanjing is a national important scientific research and education base and a comprehensive transportation hub. It stands as an important gateway city for the Yangtze River Delta to drive the development of the central and western regions. It is also one of the first batch of national historical and cultural cities and is a crucial intersection between the eastern coastal economic belt and the Yangtze economic belt. Nanjing is one of the first cities in China to be designated as a National Historic and Cultural City and is also a key scenic tourist cities in China. From the ancient charm of the Qinhuai River and the remnants of the Ming Palace to the bustling modern shopping haven of Xinjiekou, every corner of Nanjing tells a story. The city's plane is "long in the north and south narrow in the east and west", with a maximum horizontal distance of about 70 km from east to west and a maximum vertical distance of about 150 km from north to south. As of the end of 2022, Nanjing has a total area of 6587.02 km² and has a resident population of 9.4911 million.

中国 合肥

Hefei, China

　　合肥是安徽省的省会，位于巢湖之滨，坐落在跨越大别山东北延伸的一个低马鞍上。合肥是一座有着 2000 多年历史的古城，因东淝河与南淝河均发源于此而得名，历史代表文化为庐州文化、皖江文化。合肥是一座蓬勃发展的综合性工业城市，是加工贸易转移承接地区、国家汽车及零部件出口基地、国家动漫产业发展基地和服务外包基地。合肥也拥有国内一流的研究机构和大学，如中国科学技术大学。截至 2022 年底，全市总面积 11445 km²，常住人口 963.4 万。

　　Hefei is the capital of Anhui Province, located on the shore of Chaohu Lake, on a low saddle that crosses the northeastern extension of the Dabie Mountains. Hefei is an ancient city with a history of more than 2000 years, named from the confluence of the eastern and southern Fei River. Hefei's cultural heritage is epitomized by Luzhou and Wanjiang cultures. Hefei is a thriving comprehensive industrial city, serving as a hub for processing trade transfers, a national hub for automobile and parts exports, a national hub for the development of the animation industry, and a hub for service outsourcing. Furthermore, the city is home to top-tier research institutions and universities, including the University of Science and Technology of China. As of the end of 2022. Hefei has a total area of 11445 km² and has a resident population of 9.634 million.

中国 杭州
Hangzhou, China

　　杭州是浙江省的省会，位于浙江省的西北部，坐落在杭州湾头部，是长江三角洲重要城市之一。杭州位于京杭大运河的南端，已有 2200 多年历史，具代表性的文化包括西湖文化、良渚文化、丝绸文化和茶文化。杭州被列为副省级城市，是中国主要的经济和电子商务中心，以杭州为核心的都市圈是继广州 - 深圳珠江城市群、上海 - 苏州 - 无锡 - 常州城市群和北京之后的中国第四大都市圈。杭州拥有诸多风景名胜，其中西湖无疑是最著名的景点，水光山色，风景优美，为杭州带来"人间天堂"的美誉。截至 2022 年底，全市总面积 16850 km²，常住人口 1237.6 万。

　　Hangzhou is the capital of Zhejiang Province, located in the northwest of Zhejiang Province and at the head of Hangzhou Bay. It is one of the significant cities in the Yangtze River Delta. As the southern end of the Grand Canal, Hangzhou has a history of over 2,200 years, with its representative cultures including West Lake culture, Liangzhu culture, silk culture and tea culture. Classified as a sub-provincial city, Hangzhou is a primary economic and e-commerce hub in China. The metropolitan area centered on Hangzhou is the fourth largest metropolitan area in China after the Guangzhou Shenzhen Pearl River City Cluster, the Shanghai-Suzhou-Wuxi-Changzhou City Cluster and Beijing. Hangzhou has many scenic spots, among which West Lake is undoubtedly the most famous one, with beautiful water and mountains, bringing Hangzhou the reputation of "paradise on earth". As of the end of 2022, Hangzhou has a total area of 16850 km² and has a resident population of 12.376 million.

km

0   1   2   3   4

中国 南昌

Nanchang, China

　　南昌是江西省的省会，是新中国航空工业的发源地，中国重要的综合交通枢纽和光电产业基地。它位于江西省中北部，坐落于鄱阳湖平原的腹地，西邻九岭山脉，东濒鄱阳湖，连接着繁荣的华东和华南。南昌是中国重要的制造中心，制造了新中国第一架飞机、第一枚海防导弹、第一辆摩托车和拖拉机。南昌城始建于公元前 202 年，寓意"昌大南疆、南方昌盛"，是国家历史文化名城之一。2000 多年的建城史，拥有海昏侯墓、滕王阁和秋水广场等景区。截至 2022 年底，全市总面积 7195 km²，常住人口 653.81 万。

Nanchang is the capital of Jiangxi Province and the birthplace of People's Republic of China's aviation industry, an important comprehensive transportation hub and photoelectric industry base in China. Located in the north-central part of the province and in the hinterland of Poyang Lake Plain, it is bounded on the west by the Jiuling Mountains and east by the Poyang Lake, serving as a bridge connecting the prosperous East and South China. As an important manufacturing center, Nanchang produced the first aircraft, the first batch of coastal defense missiles, the first motorcycle and tractors in People's Republic of China. Nanchang was founded in 202 B.C, meaning "Prosperity in the Great South and Southern Frontier, Flourishing in the South", is one of the most famous historical and cultural city in China. More than two thousand years of history of the city, leaving behind the Haihunhou tomb, Teng Wang Pavilion, Qiushui Square and other famous tourist attractions. As of the end of 2022, Nanchang has a total area of 7195 km² and has a resident population of 6.54 million.

中国 长沙

Changsha, China

　　长沙是湖南省的省会，位于湖南省东北部湘江下游。长沙是长江中游地区重要城市和长江经济带重要的节点城市，是综合交通枢纽和国家物流枢纽，与株洲和湘潭一起构成长株潭城市群。长沙是首批国家历史文化名城，拥有 2400 多年的城市建设历史，其城名和城址始终保持不变。长沙有"工程机械之都"的美誉，是世界三大工程机械产业集聚地之一，形成了以工程机械、新材料为主体，汽车、电子信息、家电、生物医药为辅的产业链条。截至 2022 年底，全市总面积 11819 km²，常住人口 1042.06 万。

Changsha is the capital of Hunan Province, located in the lower reaches of the Xiang River in the northeastern part of Hunan Province. It serves as a key city in the middle reaches of the Yangtze River and an essential node in the Yangtze River Economic Belt. Changsha is a comprehensive transportation hub and a national logistics nexus, forming the Changsha-Zhuzhou-Xiangtan urban agglomeration alongside Zhuzhou and Xiangtan. Designated as one of the first National Historical and Cultural Cities, Changsha has a history of more than 2,400 years of urban construction, with its city name and location remaining unchanged. Known as the Capital of Engineering Machinery, Changsha is one of the three major global hubs for the construction machinery industry, forming an industrial chain with engineering machinery and new materials as the main body, supplemented by automobiles, electronic information, home appliances, biomedicine and other industries. As of the end of 2022, Changsha has a total area of 11819 km² and has a resident population of 10.4206 million.

中国 福州

Fuzhou, China

　　福州是福建省的省会，位于福建省东部、闽江下游，是福建省的政治、经济、文化中心和闽江流域的工业中心。福州是国家历史文化名城，福州的马尾区是中国近代海军的摇篮、中国船政文化的发祥地。福州的三坊七巷是福建省非常古老的文化街区，距今已有 2000 多年的历史，历经晋、唐、宋、元、明、清等历史朝代，一直是福州最具特色的文化符号之一，因其古老的文化底蕴和独特的城市格局，被誉为"里坊制度活化石"，每年吸引大量游客前来观光游览。截至 2022 年底，全市总面积 11968.53 km²，常住人口 844.8 万。

Fuzhou is the capital of Fujian Province, located in the eastern part of Fujian, at the lower reaches of the Min River. It serves as the political, economic, cultural and industrial center of Fujian Province on the Min River. Fuzhou is recognized as a National Historic and Cultural City. Mawei District in Fuzhou is heralded as the cradle of China's modern navy and the birthplace of China's shipbuilding culture. Fuzhou's Three Lanes and Seven Alleys is one of the oldest cultural district in Fujian Province, with a history of more than 2000 years. It has gone through the historical dynasties of Jin, Tang, Song, Yuan, Ming, Qing, etc, and has always been one of the most distinctive cultural symbols of Fuzhou with its ancient cultural heritage and unique urban pattern. It is known as the "living fossil of the Li Fang" system and attracts a large number of tourists every year. As of the end of 2022, Fuzhou has a total area of 11968.53 km² and has a resident population of 8.448 million.

中国 台北
Taipei, China

　　台北是台湾省的省会，位于台湾岛北部的台北盆地，四周被新北市环绕，西临淡水河及其支流新店溪，东至南港附近，南至木栅以南丘陵区，北靠大屯山东南麓。台北市是台湾省的经济、文化、工业和商业中心。全市下辖 12 个区，面积 271.8 km²。台北也是台湾省的旅游中心，迷人的自然风光、丰富的历史遗迹、多姿多彩的民族风情吸引着世界各地的游客。以台北为中心与周边市镇所联结而成台北都会区，是台湾人口最多的都会区。截至 2022 年 2 月，台北人口为 2504597 人。

Taipei is the provincial capital of Taiwan Province, located in the Taipei Basin in the northern part of Taiwan Island. It is surrounded by New Taipei City, with the Danshui River and its tributary Xindian River to the west, Nangang in the east, hilly area south of Mushan in the south, and the southeastern foothills of Datun Mountain in the north. Taipei is the economic, cultural, industrial and commercial center of the whole island. The city has 12 districts under it's jurisdiction, with an area of 271.8 km². Taipei is also the tourist center of Taiwan Province. Its charming natural scenery, rich historical relics and colorful ethnic customs attract tourists from all over the world. The Taipei Metropolitan Area, centered around Taipei and incorporating neighboring cities and towns, is the most populous metropolitan area in Taiwan Province. As of February 2022, Taipei has a population of 2504597.

中国 广州

Guangzhou, China

广州是广东省的省会，是广州都市圈核心城市，国务院批复确定的中国重要的中心城市、国际商贸中心和综合交通枢纽。广州地处中国华南地区、珠江下游、濒临南海，隔海与香港特别行政区、澳门特别行政区相望。广州是重要的港口和交通枢纽，也是海上丝绸之路的起点之一，被誉为"千年商都"。广州位于粤港澳大湾区的中心，是亚太地区重要的金融中心，行政上具有副省级地位。广州拥有 2200 多年的悠久历史，是首批国家历史文化名城。截至 2022 年底，全市总面积为 7434.40 km²，常住人口 1873.41 万。

Guangzhou is the capital of Guangdong Province, the core city of the Guangzhou metropolitan area. It is designated by the State Council as one of China's key central cities, an international business center and comprehensive transportation hub in China. Located in South China, along the lower reaches of the Pearl River and adjacent to the South China Sea, Guangzhou faces the special administrative regions of Hong Kong and Macao across the sea. Guangzhou is an important port and transportation hub, and one of the starting points of the Maritime Silk Road. It is known as the "Millennium Commercial Capital". Located in the center of the Guangdong-HongKong-Macao Greater Bay Area, it is an important financial center in the Asia-Pacific region and has sub provincial status administratively. With a long history of more than 2200 years, Guangzhou is one of the first batch of national famous historical and cultural cities. As of the end of 2022, Guangzhou has a total area of 7434.40 km² and has a resident population of 18.7341 million.

km

0   2   4   6   8

中国 深圳
Shenzhen, China

　　深圳是中国主要的副省级城市和经济特区之一，位于广东省南部、珠江口东岸，南接香港，北接东莞，东北与惠州相邻。1979年，深圳市成立，1980年，成为中国设立的第一个经济特区，中国改革开放的窗口和新兴移民城市，创造了举世瞩目的"深圳速度"。深圳是全球科技、研究、制造、商业、经济、金融、旅游和运输的中心，深圳港是世界上第四大繁忙的集装箱港口。深圳被归类为大港口城市，属于世界上最大的港口城市类型。截至2022年末，全市总面积1997.47 km²，常住人口1766.18万。

　　Shenzhen is a major sub-provincial city and one of the special economic zones of China, located in the southern part of Guangdong Province, on the eastern bank of the Pearl River estuary. It borders Hong Kong to the south, Dongguan to the north, and Huizhou to the northeast. In 1979, the city of Shenzhen was established, and in 1980, it became the first special economic zone set up in China, serving as a window to China's reform and opening up and a burgeoning immigrant city, creating the globally acclaimed "Shenzhen Speed". Shenzhen is a global center in technology, research, manufacturing, business, economics, finance, tourism and transportation, and the Port of Shenzhen is the world's fourth busiest container port. Shenzhen is classified as a Large-Port Megacity, the largest type of port-city in the world. As of the end of 2022, Shenzhen has a total area of 1997.47 km² and has a resident population of 17.6618 million.

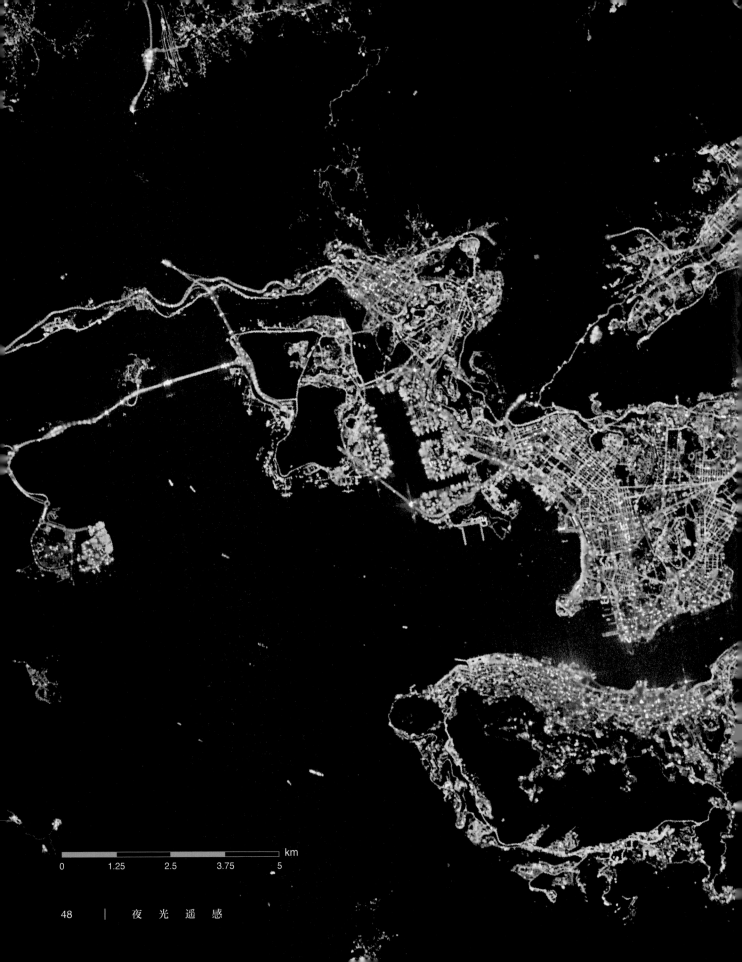

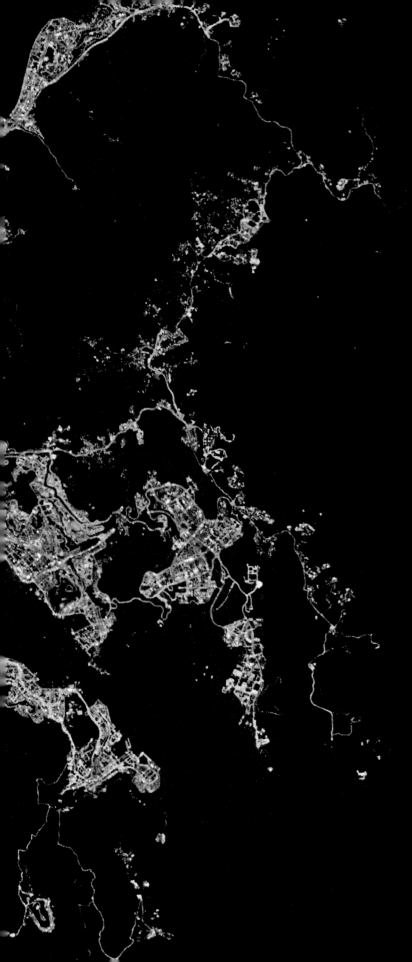

中国 香港
Hongkong, China

　　香港特别行政区位于中国南部、珠江口以东，北与深圳相邻，西与澳门隔海相望，由香港岛、九龙半岛、新界和离岛四个部分组成，陆地面积 1113.76 km²，海域面积 1641.21 km²，总面积 2754.97 km²。香港是经济、娱乐和购物的中心，其中尖沙咀、油麻地和旺角最受欢迎，新界及离岛是享受宁静假期的理想目的地。香港是一座高度繁荣的自由港和国际大都市，与纽约、伦敦齐名，是全球第三大金融中心，重要的国际贸易、航运中心和国际创新科技中心。截至 2022 年底，总人口 733.32 万人，是世界上人口密度最高的地区之一，人均寿命全球第一，人类发展指数全球第四。

The Hong Kong Special Administrative Region is located in the south of China, east of the Pearl River Estuary, borded by Shenzhen to the north and overlooking Macau to the west. It consists of Hong Kong Island, Kowloon Peninsula, the New Territories and the Outlying Islands. It has a land area of 1113.76 km², a sea area of 1641.21 km², and a total area of 2754.97 km². Hong Kong is the center of economy, entertainment and shopping, among which Tsim Sha Tsui, Yau Ma Tei and Mong Kok are the most popular, while New Territories and Islands District are ideal places to enjoy a quiet holiday. Hong Kong is a thriving free port and a global metropolis, standing alongside New York and London and is recognized as the third-largest financial center in the world. It's a significant hub for international trade, shipping, and innovative technology. As of the end of 2022, the total population is 7.3332 million, making it one of the regions with the highest population density in the world. Its average life expectancy ranks first in the world, and its human development index ranks fourth in the world.

km
0   1.25   2.5   3.75   5

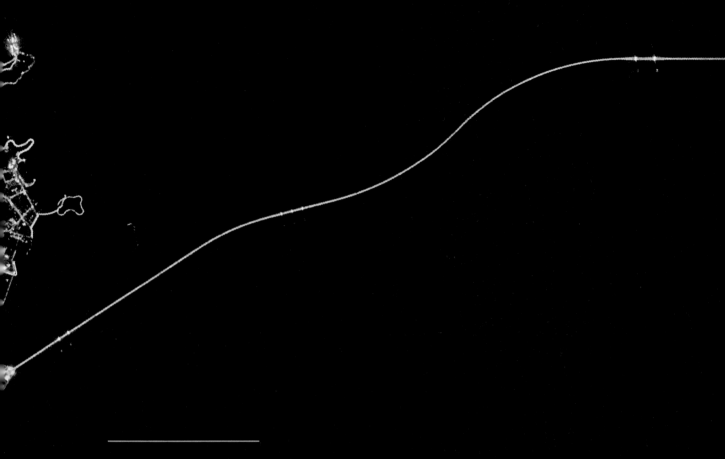

中国 澳门
Macau, China

澳门特别行政区位于中国南部、珠江口西侧，是中国大陆与中国南海的水陆交会处，与广东省珠海市相邻。澳门特别行政区由澳门半岛和凼仔、路环二岛以及路凼城（路凼填海区）组成，陆地面积 32.9 km^2。在"一国两制"的原则下，澳门是国际自由港、世界旅游休闲中心、世界四大赌城之一，也是全球人口密度最高的地区之一。其轻工业、旅游业、酒店业和娱乐场使澳门长盛不衰，成为全球发达、富裕的地区之一，同时也是度假和博彩旅游的首选目的地。截至 2022 年，澳门特别行政区常住人口 67.28 万。

The Macau Special Administrative Region is located in the south of China, wast of the Pearl River estuary, where China's mainland meets the South China Sea, and it borders Zhuhai City in Guangdong Province. The Macau Special Administrative Region consists of the Macau Peninsula, Taipa, Coloane Island and Cotai City (Cotai Reclamation Area), with a land area of 32.9 km^2. Under the principle of "one country two systems", the Macao Special Administrative Region is an international free port, a world tourism and leisure center, and one of the world's four major casinos. It is also one of the most densely populated regions in the world. Light industry, tourism, hotel industry and casinos have enabled Macao to continue to prosper and become one of the most developed and wealthy regions in the world. It has become a major resort city and a preferred destination for gambling tourism. As of 2022, the resident population of the Macao Special Administrative Region is 672.8 thousand.

中国 海口
Haikou, China

　　海口是海南省的省会，是中国"一带一路"倡议支点城市和海南自由贸易港核心城市，是海南省的政治、经济、科技和文化中心。海口位于海南岛北部，东邻文昌，西接澄迈，南毗定安，北濒琼州海峡。地处热带，是一座富有海滨自然旖旎风光的南方滨海城市。自北宋开埠以来，已有上千年的历史。海口位于海南岛的北部，并包括离岛海甸岛和新埠岛，总面积 3126.83 km²，其中，陆地面积 2296.82 km²，海域面积 830 km²。有府城鼓楼、西天庙五公祠、秀英炮台等主要景点。截至 2022 年底，海口市常住人口 293.97 万。

　　Haikou is the capital of Hainan Province, a strategic fulcrum city of the country's "Belt and Road Initiative" and the core city of Hainan Free Trade Port, as well as the political, economic, technological and cultural center of Hainan Province. Haikou is located in the northern part of Hainan Island, adjacent to Wenchang in the east, Chengmai in the west, Ding an in the south, and Qiongzhou Strait in the north. It is located in the tropics and is a southern coastal city rich in natural and beautiful seaside scenery. Since the opening of the port in the Northern Song Dynasty, it has a history of thousands of years. Haikou consists of the main island of Hainan Island, the outlying islands of Haidian Island and Xinbu Island, with a total area of 3126.83 km², of which the landarea is 2296.82 km² and the sea area is 830 km². There are Fucheng Drum Tower, Xitian Temple, Wugong Temple, Xiuying Fort and other major attractions. As of the end of 2022, Haikou City has a resident population of 2.9397 million.

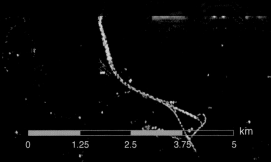

				km
0	1.25	2.5	3.75	5

中国 呼和浩特
Hohhot, China

　　呼和浩特是内蒙古自治区的首府，位于内蒙古中南部，北靠大青山，南临河套平原，是内蒙古主要的工业中心，也是推行的"西部大开发"战略的核心目标地区。呼和浩特有许多知名企业，其中包括著名的乳制品生产商——内蒙古伊利实业集团和中国蒙牛乳业有限公司。呼和浩特的街道上充满了少数民族元素。呼和浩特拥有格根塔拉草原和希拉穆仁草原的壮丽美景，也有诸如大昭寺、五塔寺和西里图召宫等奇特的文化遗址。截至 2022 年末，全市总面积 17200 km²，常住人口 355.11 万。

　　Hohhot is the capital of Inner Mongolia Autonomous Region in northern China, located in the central and southern part of Inner Mongolia, bordering the Daqing Mountain to the north and the Hetao Plateau to the south. It is the main industrial center of inner Mongolia, and also the core target area of the "Western Development" strategy implemented by the central government. Hohhot has many well-known enterprises, including famous dairy producers - Inner Mongolia Yili Industrial Group and China Mengniu Dairy Co., Ltd. Hohhot's streets are full of ethnic minority elements. Hohhot has the magnificent scenery of Gegentala Grassland and Xilamuren Grasslandas, as well as the unique cultural relics including Dazhao Temple, Wuta Temple and Xilituzhao Palace. As of the end of 2022, Hohhot has a total area of 17200 km² and has a resident population of 3.5511 million.

km
0　　1　　2　　3　　4

中国 太原
Taiyuan, China

　　太原是山西省的省会，被誉为中国北方的煤铁城。太原是山西省的政治、经济和文化中心，世界晋商都会，是以能源和重化工为主的工业城市。太原位于山西省中部，汾河从市区流过。太原是一座历史悠久的古城，始建于公元前497年，有着2500多年的建城史。太原地理位置优越，曾是中国北方的军事要塞，有中原北门之称。它也是赵国、前秦、东魏、北齐、北晋、后唐、后晋、后汉、北汉等历代王朝的都城或次都。截至2022年底，全市总面积6988 km²，常住人口为543.50万。

　　Taiyuan is the capital of Shanxi Province, known as the coal and iron city of northern China. Taiyuan is the political, economic and cultural center of Shanxi Province and is a global gathering place for Jin merchants. Predominantly an industrial city, Taiyuan's economy is driven by energy and heavy chemical industry. Taiyuan is located in the central part of Shanxi Province, with the Fen River flowing through the city. Taiyuan is an ancient city with a long history, dating back to 497 BC, with more than 2500 years of city-building history. Taiyuan has a superior geographical location and a rich historical and cultural heritage. It was once a military fortress in northern China, known as the north gate of the Central Plains. It was also the capital or secondary capital of various dynasties such as Zhao, Former Qin, Eastern Wei, Northern Qi, Northern Jin, Later Tang, Later Jin, Later Han, Northern Han and so on. As of the end of 2022, Taiyuan has a total area of 6988 km² and has a resident population of 5.435 million.

中国 乌鲁木齐

Urumqi, China

　　乌鲁木齐是新疆维吾尔自治区的首府，是中国西北地区重要的中心城市和面向中亚西亚的国际商贸中心。乌鲁木齐是中国西北内陆仅次于西安的第二大城市，地处中国西北地区、新疆中部，毗邻中亚各国，是新疆的政治、经济、文化、科教和交通中心。乌鲁木齐是世界上距离海洋最远的城市，有"亚心之都"的称号，是第二座亚欧大陆桥中国西部桥头堡和中国向西开放的重要门户。截至 2022 年底，全市总面积 13800 km²，常住人口 408.24 万。

Urumqi, the capital of the Xinjiang Uygur Autonomous Region, is an important central city in Northwest China and an international business center for Central Asia and West Asia approved by the State Council. Urumqi is the second largest city in northwest China after Xian. Urumqi City is located in Northwest China, central Xinjiang, adjacent to Central Asian countries, and is the political, economic, cultural, scientific, educational and transportation center of Xinjiang. Urumqi is the city farthest from the sea in the world and is known as the "City of Asia". It is also the western gateway of the second Eurasian Continental Bridge in China. As of the end of 2022, Urumqi, has a total area of 13800 km² and has a resident population of 4.0824 million.

km
0   1.25   2.5   3.75   5

中国 银川

Yinchuan, China

　　银川是宁夏回族自治区的首府，是中国西北地区重要的中心城市。银川位于黄河之西，贺兰山之东，东与吴忠市接壤，西依贺兰山与内蒙古自治区阿拉善盟为邻，南与吴忠市相连，北接石嘴山市。银川是古丝绸之路商贸重镇，是宁夏的经济、文化、科研、交通和金融中心。银川是中国重要的商品粮生产基地，盛产优质葡萄，羊绒产业发达。大量的西夏遗迹、壮观的贺兰山风光、引人入胜的岩画，为银川吸引了大量游客。截至 2022 年底，全市总面积 9025.38 km²，常住人口 289.68 万。

Yinchuan is the capital of Ningxia Hui Autonomous Region and an important central city in Northwest China. Yinchuan is located in the west of the Yellow River, east of Helan Mountain, bordering Wuzhong City in the east, adjacent to Alxa League of Inner Mongolia Autonomous Region in the west by Helan Mountain, connected with Wuzhong City in the south, and Shizuishan City in the north. Yinchuan is an important commercial and trade center on the ancient Silk Road, and the economic, cultural, scientific research, transportation and financial center of Ningxia. Yinchuan is an important commercial grain production base in China, rich in high-quality grapes and a well-developed cashmere industry. A large number of Xi Xia Dynasty relics, the spectacular Helan Mountain scenery and fascinating rock paintings have attracted a large number of tourists. As of the end of 2022, Yinchuan has a total area of 9025.38 km² and has a resident population of 2.8968 million.

中国 西宁
Xining, China

　　西宁是青海省的省会，位于湟水河谷，是青藏高原最大的城市及其东方门户。这座历史悠久的城市曾是"丝绸之路"南路和"唐蕃古道"的必经之地，自古以来便是西北地区的交通要道和军事重地。长达 2000 多年的历史中，西宁在北方丝绸之路河西走廊上始终扮演着商业中心的角色，并多次作为古代中央王朝防范西方游牧民族入侵的军事前线。这里拥有许多对穆斯林和佛教徒都具有重要宗教意义的遗址，包括东莞清真寺和塔尔寺。西宁现在正深度融入"一带一路"建设，努力将兰州—西宁城市群发展为维护国土安全、生态安全及西北地区繁荣稳定的重要城市群。截至 2022 年底，全市总面积 7660 km^2，常住人口 248 万。

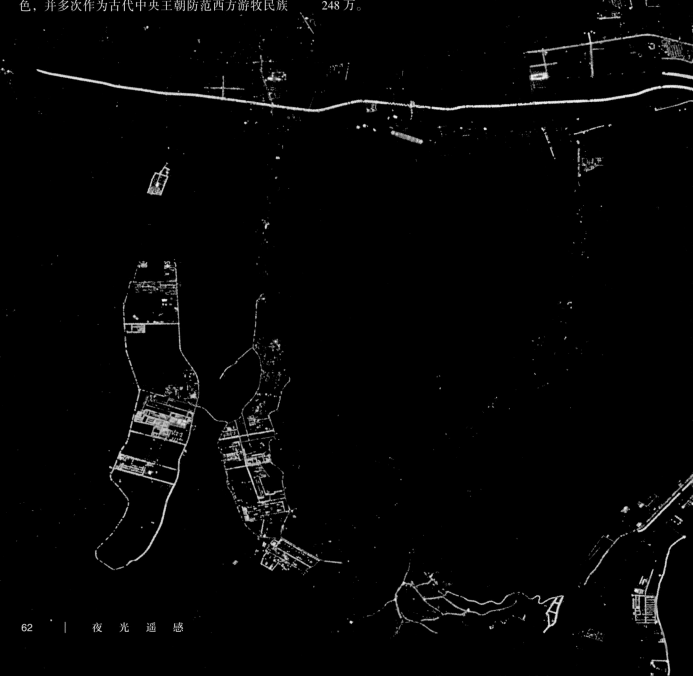

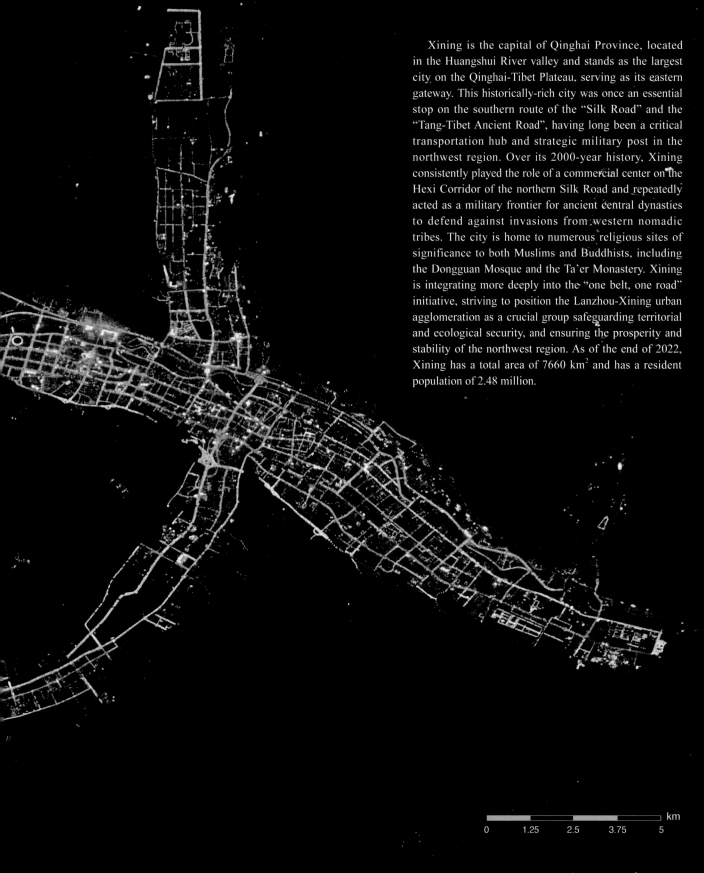

Xining is the capital of Qinghai Province, located in the Huangshui River valley and stands as the largest city on the Qinghai-Tibet Plateau, serving as its eastern gateway. This historically-rich city was once an essential stop on the southern route of the "Silk Road" and the "Tang-Tibet Ancient Road", having long been a critical transportation hub and strategic military post in the northwest region. Over its 2000-year history, Xining consistently played the role of a commercial center on the Hexi Corridor of the northern Silk Road and repeatedly acted as a military frontier for ancient central dynasties to defend against invasions from western nomadic tribes. The city is home to numerous religious sites of significance to both Muslims and Buddhists, including the Dongguan Mosque and the Ta'er Monastery. Xining is integrating more deeply into the "one belt, one road" initiative, striving to position the Lanzhou-Xining urban agglomeration as a crucial group safeguarding territorial and ecological security, and ensuring the prosperity and stability of the northwest region. As of the end of 2022, Xining has a total area of 7660 km^2 and has a resident population of 2.48 million.

km

| 0 | 1.25 | 2.5 | 3.75 | 5 |

中国 兰州
Lanzhou, China

　　兰州是甘肃省的省会，是西部地区重要的中心城市和丝绸之路经济带的重要节点，是国务院批复确定的中国西北地区重要的工业基地和综合交通枢纽。兰州地处中国西北地区、中国陆域版图的几何中心，地势西南高，东北低，黄河从西南向东北蜿蜒流淌，峡谷与盆地交错分布。兰州因丝绸之路的繁盛而显赫，逐渐崭露为重要的交通要道和商埠重镇。后来，该市成为中国最早的近代工业化城市之一。中华人民共和国成立后，兰州被定位为国家的主要工业基地，发展为重要的石油化工、生物制药和装备制造基地。截至 2022 年底，全市总面积 13100 km^2，常住人口 441.53 万。

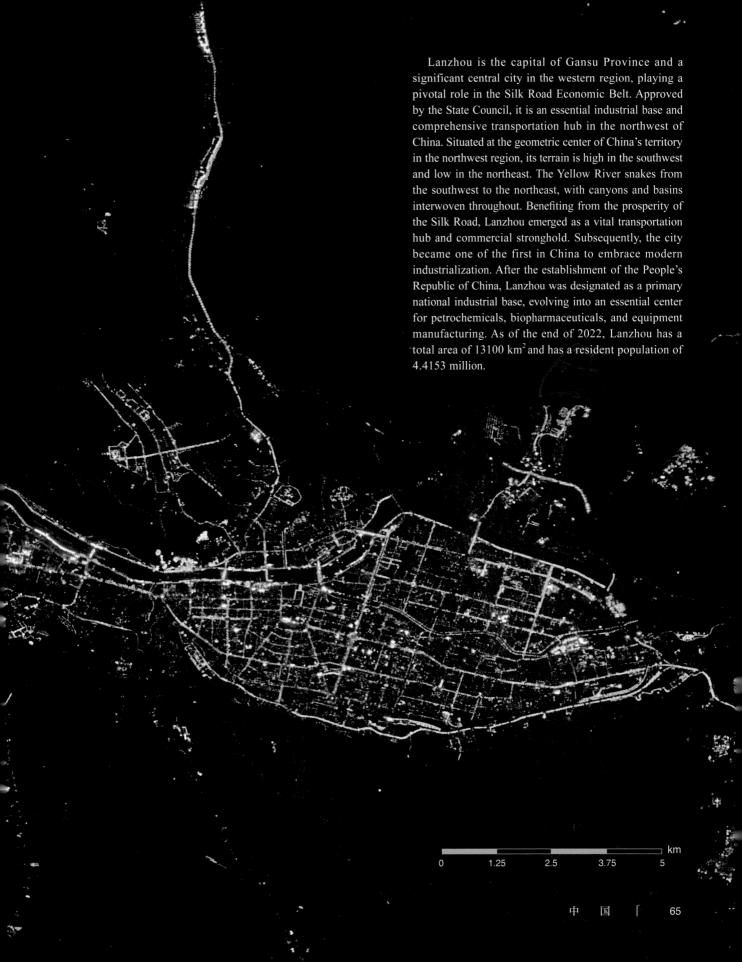

Lanzhou is the capital of Gansu Province and a significant central city in the western region, playing a pivotal role in the Silk Road Economic Belt. Approved by the State Council, it is an essential industrial base and comprehensive transportation hub in the northwest of China. Situated at the geometric center of China's territory in the northwest region, its terrain is high in the southwest and low in the northeast. The Yellow River snakes from the southwest to the northeast, with canyons and basins interwoven throughout. Benefiting from the prosperity of the Silk Road, Lanzhou emerged as a vital transportation hub and commercial stronghold. Subsequently, the city became one of the first in China to embrace modern industrialization. After the establishment of the People's Republic of China, Lanzhou was designated as a primary national industrial base, evolving into an essential center for petrochemicals, biopharmaceuticals, and equipment manufacturing. As of the end of 2022, Lanzhou has a total area of 13100 km^2 and has a resident population of 4.4153 million.

km
0   1.25   2.5   3.75   5

中国 西安

Xi'an, China

　　西安是陕西省的省会，是西安都市圈以及关中平原城市群核心城市之一。自 20 世纪 80 年代起，随着中国内陆尤其是中西部地区的经济增长，西安逐渐成为中西部地区的文化、工业、政治和教育中心，并设有众多研发机构。西安是国务院批复确定的中国西部重要的中心城市，也是国家重要的科研、教育和工业基地。西安是中国最受欢迎的旅游目的地之一，是丝绸之路的起点，也是秦始皇兵马俑的所在地，这两处遗址均被联合国教科文组织列为世界历史文化遗产。截至 2022 年底，全市总面积 10108 km²，常住人口 1299.59 万人。

Xi'an is the capital of Shaanxi Province and one of the core cities of the Xi'an metropolitan area and the Guanzhong Plain urban agglomeration. Since the 1980s, with the economic growth of China's inland and especially the central and western regions, Xi'an has gradually become the cultural, industrial, political and educational center of the central and western regions, and has many research and development institutions. Xi'an is an important central city in western China approved by the State Council, and also an important national scientific research, education and industrial base. Xi'an is one of the most popular tourist destinations in China, as it is the starting point of the Silk Road and the location of the Terracotta Army of Qin Shi Huang. These two sites are both listed as World Heritage Sites by UNESCO. As of the end of 2022, Xi'an has a total area of 10108 km² and has a resident population of 12.9959 million.

中国 重庆
Chongqing, China

重庆是中国的四大直辖市之一，是中国重要的现代化制造业基地、金融中心和国际交通枢纽。从地理位置上看，重庆是通往中国西部的门户，是长江经济带的重要枢纽，是中国"一带一路"倡议的战略基地，有世界文化遗产大足石刻、世界自然遗产武隆喀斯特和南川金佛山等多处世界级景观。在行政上，它是中央政府直接管理的四个直辖市之一，也是唯一一个位于内陆的直辖市。重庆是中国唯一一个常住人口超过 3000 万的城市。截至 2022 年底，重庆总面积 82402 km²，常住人口 3213.34 万。

Chongqing is one of the four municipalities directly under the central government in China. It is an important modern manufacturing base, financial center and international transportation hub in China. Geographically, Chongqing is the gateway to the west of China, an important hub of the Yangtze River Economic Belt, and a strategic base of China's "One Belt, One Road" initiative. There are world cultural heritage Dazu Rock Carvings, world natural heritage Wulong Karst and Nanchuan Jin Foshan and other landscapes. Administratively, it is one of four municipalities directly administered by the central government, and the only one located inland. Chongqing is the only city in China with a permanent population of over 30 million. As of the end of 2022, Chongqing has a total area of 82402 km² and has a resident population of 3.21334 million.

km
0   2   4   6   8

中国 成都
Chengdu, China

　　成都是四川省的省会，是一个副省级城市，它是除四个直辖市外唯一一个人口超过 2000 万的城市。成都位于四川中部，历来是西部地区的重要枢纽。成都是中国最重要的经济、金融、商业、文化、交通和通信中心之一，以机械、汽车、医药、食品和信息技术产业为特色。成都市是国务院批复确定的国家高新技术产业基地、商贸物流中心、综合交通枢纽，以及西部地区的核心城市，更是成渝经济圈的核心城市。这座古老的城市自古就有"天府之国"的美誉，是首批国家历史文化名城。截至 2022 年底，成都全市总面积 14335 km²，常住人口 2126.8 万。

　　Chengdu, the capital of Sichuan Province, is a sub-provincial city, and it is the only city with a population of more than 20 million besides the four municipalities directly under the Central Government. Located in central Sichuan, Chengdu has traditionally been a hub for western China. Chengdu is one of the most important economic, financial, commercial, cultural, transportation and communication centers in China, featuring machinery, automobile, medicine, food and information technology industries. Chengdu is an important national high-tech industrial base, commercial logistics center and comprehensive transportation hub approved by the State Council, an important central city in the western region, and a core city in the Chengdu-Chongqing economic circle. Chengdu has been known as the "Land of Abundance" since ancient times, and is one of the first batch of national historical and cultural cities. As of the end of 2022, Chengdu has a total area of 14335 km² and has a resident population is 21.268 million.

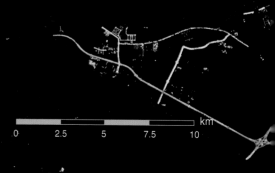

km
0　　2.5　　5　　7.5　　10

中国 拉萨

Lhasa, China

拉萨是西藏自治区的行政首府，是中国具有雪域高原风情和浓郁民族特色的国际旅游城市。拉萨市地处中国西南地区、西藏高原中部、喜马拉雅山脉北麓和雅鲁藏布江支流拉萨河的中游河谷。不仅是西藏的经济、文化和科教中心，也是藏传佛教的圣地，包含众多具有重要历史和宗教意义的藏传佛教遗址，如世界著名的布达拉宫、大昭寺和罗布林卡宫。拉萨海拔 3650 m，是世界上海拔最高的城市之一，全年多晴朗天气，素有"日光城"的美誉。截至 2022 年底，全市面积 29640 km²，户籍人口 58.12 万人。

Lhasa, the administrative capital of the Tibet Autonomous Region, is an international tourist city with snow-covered plateaus and ethnic characteristics. Lhasa City is located in the southwest of China, the central part of the Tibetan Plateau, the northern side of the Himalayas, and the valley plain of the middle reaches of the Lhasa River, a tributary of the Yarlung Zangbo River. It is not only the economic, cultural, scientific and educational center of Tibet, but also the holy land of Tibeton Buddhit. Tibetan Buddhist sites such as Potala Palace, Jokhang Temple and Norbulingka Palace. At an altitude of 3650 m, Lhasa is one of the cities with the highest altitudes in the world. It has sunny weather throughout the year and is known as the "Sunshine City". As of the end of 2022, Lhasa has a total area of 29640 km² and has a registered population of 581200.

中国 贵阳

Guiyang, China

　　贵阳是贵州省的省会，地处黔中山原丘陵中部，海拔约 1100 m，是首个国家森林城市和重要的生态休闲度假城市。作为贵州的政治、经济、文化、科教和交通中心，贵阳是西南地区重要的交通和通信枢纽、工业基地及商贸旅游服务中心，同时也是国家大数据产业发展集聚地和国家大数据综合试验区的核心。贵阳历史上以铝生产、磷矿开采和光学仪器制造为主，经过改革后，服务业逐渐成为其经济增长的主要动力。截至 2022 年底，全市总面积 8043 km²，常住人口 622.04 万人，地区生产总值 4921.17 亿元。

Guiyang, the capital of Guizhou Province, is located in the middle of the hills in the middle of Guizhou, with an altitude of about 1100 m. It is the first national forest city and an important ecological leisure and tourism city. Guiyang is the center of politics, economy, culture, science and education, and transportation in Guizhou Province. It is an important transportation and communication hub, industrial base, and business and tourism service center in Southwest China. It is also the core of the national big data industry development cluster and the national big data comprehensive test area. In its history, Guiyang was primarily known for aluminum production, phosphate mining, and the manufacturing of optical instruments. After undergoing reforms, the service industry gradually emerged as the main driving force behind its economic growth. As of the end of 2022, the city's total area is 8043 km², with a permanent population of 6.2204 million and a regional GDP of 492.117 billion yuan.

km

0　　2.5　　5　　7.5　　10

0   2.5   5   7.5   10   km

中国 昆明
Kunming, China

　　昆明是云南省的省会，是云南省的政治、经济、交通和文化中心，位于东盟"10+1"自由贸易区经济圈、大湄公河次区域经济合作圈和泛珠三角区域经济合作圈的交汇点。昆明地处云贵高原中部，海拔1900 m，纬度在北回归线以北，周围环绕着寺庙、湖泊和独特的石灰岩山丘。得天独厚的地理位置为昆明赋予了北亚热带低纬高原山地季风气候，使其四季温暖如春，被誉为"春城"。昆明与多个东南亚国家接壤，与越南有铁路相连，与缅甸、老挝和泰国则通过公路相接，使得它成为通往东南亚和南亚的门户和重要的商业贸易中心。截至2022年底，昆明总面积21012.54 km²，常住人口860万人，地区生产总值7541.37亿元。

　　Kunming is the capital of Yunnan Province and the political, economic, transportation and cultural center of Yunnan Province. It is located at the intersection of the ASEAN "10+1" Free Trade Zone Economic Circle, the Greater Mekong Sub-regional Economic Cooperation Circle, and the intersection of Pan-Pearl River Delta Regional Economic Cooperation Circle. Kunming is located in the middle of the Yunnan-Guizhou Plateau, 1900 m above sea level, and the latitude is north of the Tropic of Cancer, surrounded by temples and lakes and limestone hills. Kunming has a north subtropical low-latitude plateau mountainous monsoon climate. It is a mountainous landform. Because it is located in a low-latitude plateau, it forms a "spring-like climate" and enjoys the reputation of "Spring City". Kunming borders many Southeast Asian countries and is the gateway to Southeast Asia and South Asia. It is connected to Vietnam by railway and to Myanmar, Laos and Thailand by road, making Kunming an important commercial and trade center. As of the end of 2022, Kunming has a total area of 21012.54 km², a permanent population of 8.6 million, and a regional GDP of 754.137 billion yuan.

中国 南宁

Nanning, China

　　南宁是广西壮族自治区的首府，位于广西南部，是该地区的经济、金融和文化中心，是"一带一路"的重要节点，是北部湾经济区、珠江——西江经济带和北部湾城市群核心城市。其独特的地理位置，让它成为华南、西南和中国-东盟经济圈的交汇之地。南宁近海、近边、沿江，展现出"两近两边"的特点。素有"中国绿城"的美誉，先后荣获联合国人居奖、全国文明城市、国家生态园林城市、国家卫生城市、国家森林城市和中国优秀旅游城市等荣誉称号。南宁市积极发展向海经济，推动跨境产业融合发展，立足区位交通优势向枢纽经济转型，服务构建中国——东盟命运共同体。截至2022年底，全市面积22100 km²，常住人口889.17万人，地区生产总值5218.34亿元。

Nanning is the capital of Guangxi Autonomous Region. It is located in the south of Guangxi. It is the economic, financial and cultural center of Guangxi. It is an important node of the "Belt and Road". The meeting point of the South China Economic Circle, the Southwest Economic Circle and the China-ASEAN Economic Circle has the characteristics of being near the sea, near the border, along the river, and "near both sides". Known as the "Green City of China", Nanning has won the United Nations Habitat Award and the titles of National Civilized City, National Ecological Garden City, National Sanitary City, National Forest City, and China's Excellent Tourism City. Nanning actively develops the seaward economy, promotes the integrated development of cross-border industries, transforms into a hub economy based on its location and transportation advantages, and serves to build a China-ASEAN community of shared future. By the end of 2022, the city has an area of 22100 km², a permanent population of 8.8917 million, and a regional GDP of 521.834 billion yuan.

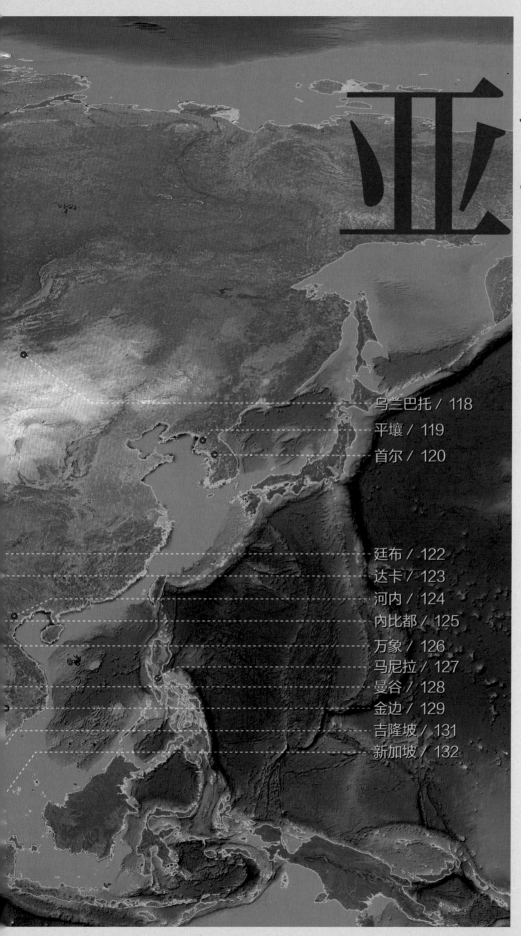

亚洲

ASIA

哈萨克斯坦 阿斯塔纳

Astana, Kazakhstan

　　阿斯塔纳位于哈萨克斯坦中北部的伊希姆河畔，隶属于阿克莫拉州。阿斯塔纳于 1997 年成为哈萨克斯坦的首都，从那时起，它在经济上迅速发展壮大，并成为中亚最现代化的城市之一。近年来，阿斯塔纳逐渐成为一个备受关注的旅游目的地。2021年，哈萨克斯坦政府将其列为国家十大旅游重点发展城市。阿斯塔纳四季温度偏冷，年均温是 1.8℃，1 月均温度为 –17.3℃，而最热的月份是 7 月，月均温度达 20.2℃。

　　Astana lies on the banks of the Ishim River in the north-central part of Kazakhstan, within the Akmola Region. The city became the capital of Kazakhstan in 1997; since then it has grown and developed economically into one of the most modern cities in Central Asia. In 2021, the government selected Astana as one of the 10 priority destinations for tourist development. Generally, Astana has a cool climate and a lively ecosystem. Astana is both the national hub of industries and transport hub of the country. The average annual temperature in Astana is 1.8℃ . The coldest month in Astana is January with a monthly average temperature of −17.3℃ . The hottest month is July, when the average monthly temperature reaches 20.2℃ .

比什凯克
吉尔吉斯斯坦

km
0   1.25   2.5   3.75   5

吉尔吉斯斯坦 比什凯克
Bishkek, Kyrgyzstan

比什凯克是吉尔吉斯斯坦的首都和最大城市，坐落在哈萨克斯坦和吉尔吉斯斯坦的边境附近，海拔约 800 m。城市位于吉尔吉斯斯坦阿拉图山脉北麓，该山脉是天山山脉的延伸，海拔高度达 4895 m。在城市北部，广袤的肥沃草原向北延伸至邻国哈萨克斯坦。比什凯克拥有宽阔的林荫大道和大理石面的公共建筑，以及许多围绕着内部庭院布局的苏联时代风格公寓楼。城市的街道呈网格状，两侧大多有狭窄的灌溉渠道，为树木提供水源，并在炎热的夏天为市民提供遮荫。比什凯克的城市规划和建筑风格展现出独特的历史和文化底蕴，是吉尔吉斯斯坦的政治、经济和文化中心。

Bishkek, the capital and largest city of Kyrgyzstan, located near the border between Kazakhstan and Kyrgyzstan. Bishkek is situated at an altitude of about 800 metres, just off the northern fringe of the Kyrgyz Ala-Too Range, an extension of the Tian Shan mountain range. These mountains rise to a height of 4895 metres. North of the city, a fertile and gently undulating steppe extends far north into neighbouring Kazakhstan. The river Chvy drains most of the area. Bishkek is connected to the Turkestan–Siberia Railway by a spur line.Bishkek is a city of wide boulevards and marble-faced public buildings combined with numerous Soviet-style apartment blocks surrounding interior courtyards. Streets follow a grid pattern, with most flanked on both sides by narrow irrigation channels, which provide water to trees which provide shade during the hot summers. As the major political, economical and cultural center of Kyrgyzstan, Bishkek has a unique building and city planning styles, displaying its rich history background.

乌兹别克斯坦 塔什干

Tashkent, Uzbekistan

　　塔什干是乌兹别克斯坦的首都和最大城市,也是中亚人口最多的城市,截至 2023 年的人口约为 299.95 万。它位于乌兹别克斯坦东北部,毗邻哈萨克斯坦边境。其名"塔什干"源于突厥语的"tash"(石头)和"kent"(城市),直译为"石城"或"石头之城"。在公元 8 世纪中叶,塔什干受到了粟特和突厥文化的影响,而伊斯兰教的传播也为这座城市带来了新的文化。如今的塔什干,作为乌兹别克斯坦的首都,是以乌兹别克人为主的多民族人口城市。塔什干不仅是乌兹别克斯坦的政治、文化、经济中心,拥有现代化的设施和建筑,而且还是重要的交通枢纽,连接着乌兹别克斯坦与周边国家。

Tashkent is the capital and largest city of Uzbekistan. It is the most populous city in Central Asia, with a estimated population in 2023 of 2999500. It is in northeastern Uzbekistan, near the border with Kazakhstan. Tashkent comes from the Turkic tash and kent, literally translated as "Stone City" or "City of Stones". Before Islamic influence started in the mid-8th century AD, Tashkent was influenced by the Sogdian and Turkic cultures. Today, as the capital of an independent Uzbekistan, Tashkent retains a multiethnic population, with ethnic Uzbeks as the majority.

巴基斯坦 伊斯兰堡
Islamabad, Pakistan

　　伊斯兰堡是巴基斯坦的首都，是该国人口第九多的城市，人口超过 100 万。它由巴基斯坦政府联邦直接管辖，是伊斯兰堡首都地区的一部分。在 20 世纪 60 年代，伊斯兰堡建成，取代了历史悠久的拉瓦尔品第，成为巴基斯坦的首都。伊斯兰堡以拥有多个公园和森林而著称，其中包括马加拉山国家公园和沙卡尔帕里安公园。这里有多个地标性建筑，最为著名的是雄伟的费萨尔清真寺，它不仅是巴基斯坦的象征，还是世界上最大的清真寺之一。其他著名的地标包括巴基斯坦纪念碑和民主广场。

Islamabad is the capital city of Pakistan. It is the country's ninth-most populous city, with a population of over 1 million people, and is federally administered by the Pakistani government as part of the Islamabad Capital Territory. Built as a planned city in the 1960s, it replaced Rawalpindi as Pakistan's national capital. Islamabad is known for the presence of several parks and forests, including the Margalla Hills National Park and the Shakarparian. It is home to several landmarks, including the country's flagship Faisal Mosque, which is one of the world's largest mosques. Other prominent landmarks include the Pakistan Monument and Democracy Square.

杜尚别 塔吉克斯坦

0　1.25　2.5　3.75　5　km

塔吉克斯坦 杜尚别
Dushanbe, Tajikistan

　　杜尚别是塔吉克斯坦的首都和最大的城市。它位于吉萨尔盆地，海拔介于 750 ～ 930 m 之间，北部和东部是吉萨尔山脉。杜尚别的现代文化起源于 20 世纪 20 年代，尤其在苏联统治时期，各种艺术形式如音乐、歌剧、戏剧、雕塑和电影都在此蓬勃发展。杜尚别的建筑风格经历了从新古典主义到极简主义再到现代风格的过渡。该市贡献了塔吉克斯坦国内生产总值的 20%，拥有庞大的工业、金融、零售和旅游业。城市的公园和主要景点包括胜利公园、鲁达基公园、塔吉克斯坦国家博物馆、杜尚别旗杆和塔吉克斯坦国家文物博物馆，反映出这个城市的独特魅力和丰富历史。

　　Dushanbe is the capital and largest city of Tajikistan. Dushanbe is located in the Gissar Valley. Dushanbe's modern culture had its start in the 1920s, where Soviet music, opera, theater, sculpture, film, and sports all began. The architecture of Dushanbe, once neoclassical, transitioned to a minimalist and eventually modern style. The city makes up 20% of Tajikistan's GDP and has large industrial, financial, retail, and tourism sectors. Parks and main sights of the city include Victory Park, Rudaki Park, the Tajikistan National Museum, the Dushanbe Flagpole, and the Tajikistan National Museum of Antiquities.

阿富汗 喀布尔
Kabul, Afghanistan

喀布尔是阿富汗的首都和最大的城市,位于该国的东半部,隶属于喀布尔省,在行政上划为 22 个市辖区。在当代,喀布尔一直是阿富汗的政治、文化和经济中心,快速的城市化使其成为世界上第 75 大城市。喀布尔位于兴都库什山脉之间的一个山谷中,喀布尔河穿城而过。喀布尔海拔高达 1790 m,是世界上海拔最高的首都之一。据记载,喀布尔已有超过 3500 年的历史,至少可以追溯到阿契美尼德波斯帝国时期。喀布尔是中亚和南亚贸易路线的战略要地,也是古代丝绸之路的重要目的地,长期以来都被视为鞑靼、印度和波斯的交汇点。喀布尔以其历史悠久的花园、集市和宫殿而闻名,包括巴布尔花园和达鲁阿曼宫。

Kabul is the capital and largest city of Afghanistan. Located in the eastern half of the country, it is also a municipality, forming part of the Kabul Province; it is administratively divided into 22 municipal districts. In contemporary times, the city has served as Afghanistan's political, cultural, and economical centre, and rapid urbanisation has made Kabul the 75th-largest city in the world. The modern-day city of Kabul is located high up in a narrow valley between the Hindu Kush, and is bounded by the Kabul River. At an elevation of 1790 metres, it is one of the highest capital cities in the world. Kabul is said to be over 3500 years old, mentioned since at least the time of the Achaemenid Persian Empire. It is situated in a strategic location along the trade routes of Central Asia and South Asia, and was a key destination on the ancient Silk Road; It was traditionally seen as the meeting point between Tartary, India, and Persia. Kabul is known for its historical gardens, bazaars, and palaces; well-known examples are the Gardens of Babur and Darul Aman Palace.

阿塞拜疆　巴库

km
0　　1.25　　2.5　　3.75　　5

阿塞拜疆　巴库

Baku, Azerbaijan

　　巴库是阿塞拜疆的首都和最大的城市，同时也是里海和高加索地区最大的城市。巴库位于海平面以下 28 m，是世界上海拔最低的国家首都，也是世界上海拔最低的大城市。巴库坐落于阿布歇隆半岛的南岸，紧邻巴库湾。作为阿塞拜疆的科学、文化和工业中心，巴库近年来逐渐崭露头角，并成为国际活动的重要场所。巴库国际海上贸易港每年能够处理 200 万吨普通和干散货物，为巴库乃至整个国家的经济和贸易发展提供了重要的支持。此外，巴库也因强风而闻名，这种独特的自然现象为它赢得了"风之城"的称号。这座城市的独特地理位置和自然条件赋予了了它独特的气候和风貌。

Baku is the capital and largest city of Azerbaijan, as well as the largest city on the Caspian Sea and of the Caucasus region. Baku is located 28 m below sea level, which makes it the lowest lying national capital in the world and also the largest city in the world located below sea level. Baku lies on the southern shore of the Absheron Peninsula, alongside the Bay of Baku. The city is the scientific, cultural, and industrial center of Azerbaijan. In recent years, Baku has become an important venue for international events. The Baku International Sea Trade Port is capable of handling two million tons of general and dry bulk cargoes per year. The city is renowned for its harsh winds, which is reflected in its nickname, the "City of Winds".

伊朗 德黑兰

Tehran, Iran

　　德黑兰是伊朗的首都和德黑兰省最大的城市，人口约900万，是伊朗和西亚人口最多的城市之一。德黑兰拥有许多历史遗址，其中包括戈列斯坦宫、萨德阿巴德王宫和尼亚瓦兰宫皇家建筑群，这些地方曾是前伊朗帝国最后两个王朝的所在地。德黑兰有着许多独特建筑，比如默德塔和自然之桥。作为伊朗的首都，德黑兰不仅是政治和行政中心，还是文化、艺术和经济活动的重要枢纽。该城市属于大陆性半干旱气候，城内各地区的海拔高度不一，北部山丘地带的气候通常较南部的平原凉爽。

Tehran is the largest city in Tehran Province and the capital of Iran. With a population of around 9 million in the city, Tehran is one of the most populous cities in Iran and Western Asia. Tehran is home to many historical locations, including the royal complexes of Golestan Palace, Sa'dabad Palace, and Niavaran Palace, where the two last dynasties of the former Imperial State of Iran were seated. Tehran has many unique buildings like the Milad Tower and the Tabiat Bridge. Tehran is more than a political capital for Iran, it is also a national center of culture, arts and economy. Tehran has a semi-arid climate, with different altitudes in different parts of the city, and the climate in the northern hill region is usually cooler than in the plains to the south.

德黑兰

伊　朗

km
0　1　2　3　4

亚美尼亚

◉ 埃里温

亚美尼亚 埃里温
Yerevan, Armenia

位于赫拉兹丹河畔的埃里温是亚美尼亚的首都和国家行政、文化和工业中心，也是世界上最古老的持续有人居住的城市之一。2012 年，埃里温因其丰富的文化底蕴被联合国教科文组织授予"世界图书之都"的称号，充分肯定了这座城市对文化和知识的重视。埃里温歌剧院是亚美尼亚首都的主要表演大厅，亚美尼亚国家美术馆是亚美尼亚最大的艺术博物馆，与亚美尼亚历史博物馆共享一座宏伟的建筑。这座古老的城市以其独特的历史和文化遗产而闻名，年复一年地吸引着世界各地的游客前来探寻其千年的故事。

Yerevan is the capital and largest city of Armenia and one of the world's oldest continuously inhabited cities. Situated along the Hrazdan River, Yerevan is the administrative, cultural, and industrial center of the country. Yerevan was named the 2012 World Book Capital by UNESCO. Yerevan Opera Theatre is the main spectacle hall of the Armenian capital, the National Gallery of Armenia is the largest art museum in Armenia and shares a building with the History Museum of Armenia. Every year tourists around world visit Yerevan seeking cultural enrichment.

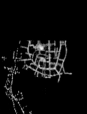

◎ 安卡拉

土耳其

土耳其 安卡拉
Ankara, Turkey

　　安卡拉是土耳其的首都，位于安纳托利亚中部，是土耳其第二大城市。这座城市的历史中心坐落在一座岩石山丘上，高出萨卡里亚河支流安卡拉河左岸 150 m。古老的安卡拉古城堡遗迹仍然屹立于此，整个城市还保留着许多罗马和奥斯曼时期的建筑，其中最引人注目的要数公元前 20 年建造的奥古斯都和罗马神庙，该神庙以安卡拉铭文而闻名，刻录在奥古斯都功德碑上。这片土地盛产梨、蜂蜜和马斯喀特葡萄。虽然安卡拉是土耳其最干旱的地区之一，但人均绿地面积达到 72 m^2，是一个绿化率较好的城市。

　　Ankara is the capital of Turkey. Located in the central part of Anatolia, the city is Turkey's second-largest city. The historical center of Ankara is a rocky hill rising 150 m over the left bank of the Ankara River, a tributary of the Sakarya River. The hill remains crowned by the ruins of Ankara Castle. There are well-preserved examples of Roman and Ottoman architecture throughout the city, the most remarkable being the 20 BC Temple of Augustus and Rome that boasts the Monumentum Ancyranum, the inscription recording the Res Gestae Divi Augusti. The area is also known for its pears, honey and Muscat grapes. Although Ankara is one of the driest areas in Türkiye, the per capita green area reaches 72 m^2, making Ankara a city with good greening rate.

伊拉克 巴格达
Baghdad, Iraq

巴格达◉

伊拉克

km
0 1 2 3 4

　　巴格达是伊拉克的首都，位于美索不达米亚平原中部地区，距幼发拉底河约 30 km，靠近巴比伦古城和萨珊王朝的古波斯首都克特西丰的遗址，底格里斯河流过巴格达市区。巴格达属副热带干燥气候，为世界上最热的城市之一，6 月至 8 月间，月平均气温最高可达 44℃，正午最高气温可达 50℃，年降雨少。公元 762 年，巴格达被选为阿拔斯王朝的首都。830 年，哈里发马蒙在巴格达创建国家学术研究机构"智慧馆"，聚集不同民族及宗教信仰的著名学者，将希腊、波斯、印度等国的古典著作加以收藏、整理并翻译成阿拉伯文，促进了阿拉伯科学文化的发展，同开罗、科尔多瓦并称为伊斯兰世界三大文化名城。

Baghdad is the capital of Iraq. It is located in the central region of the Mesopotamian Plain, with the Tigris River flowing through downtown Baghdad, while the Euphrates River is about 30 kilometres away, near the ruins of the ancient city of Babylon and the Sassanid Persian capital of Ctesiphon. Baghdad has a dry subtropical climate, making it one of the hottest cities in the world, with average monthly temperatures reaching a maximum of 44℃ between June and August, and midday highs of up to 50℃, with little annual rainfall. In 762 CE, Baghdad was chosen as the capital of the Abbasid Caliphate. In 830 CE, Caliph Mamun established the national academic research institution "House of Wisdom" in Baghdad, bringing together renowned scholars from various ethnicities and religious beliefs. They collected, organized, and translated classical works from Greece, Persia, India, and other countries into Arabic, which greatly contributed to the development of Arab scientific and cultural advancement. This effort, alongside Cairo and Cordoba, earned Baghdad the reputation of being one of the three great cultural cities in the Islamic world.

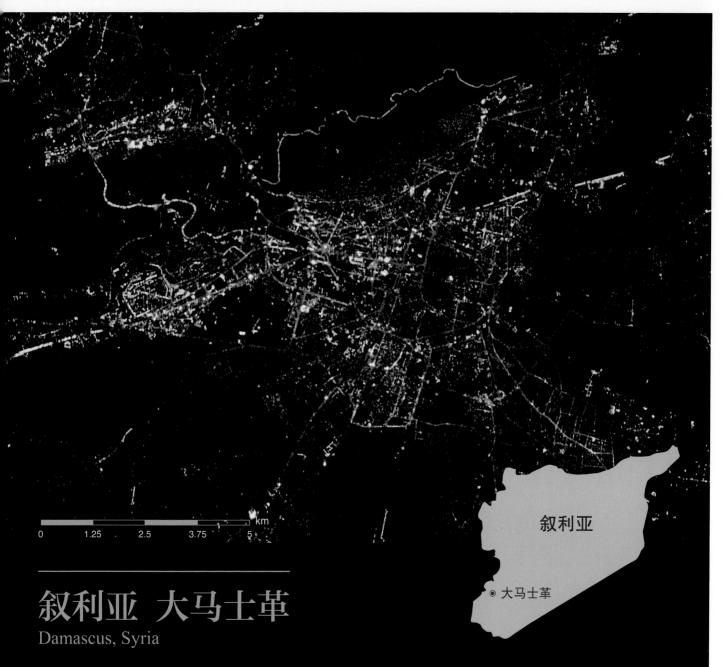

叙利亚 大马士革
Damascus, Syria

叙利亚

◎ 大马士革

　　大马士革是叙利亚的首都,世界上最古老的首都之一。在叙利亚,大马士革享有"茉莉花之城"的美名,是阿拉伯世界的文化中心之一。这座历史名城位于叙利亚西南部,是一个大都市区的中心,坐落在距地中海东岸仅 80 km 处的前黎巴嫩山脉东麓,位于海拔 680 m 的高原上,因雨影效应而气候干燥。巴拉达河流经大马士革,为这片古老的土地带来了生机。作为世界上最古老的持续有人居住的城市之一,大马士革拥有悠久的历史和文化传承。

Damascus is the capital of Syria, the oldest capital in the world and, according to some. Known colloquially in Syria as the "City of Jasmine", Damascus is one of the cultural center of the Arab world. Situated in southwestern Syria, Damascus is the center of a large metropolitan area. Nestled among the eastern foothills of the Anti-Lebanon mountain range 80 km inland from the eastern shore of the Mediterranean on a plateau 680 m above sea level, Damascus experiences a dry climate because of the rain shadow effect. The Barada River flows through Damascus. Damascus is one of the oldest continuously inhabited cities in the world.

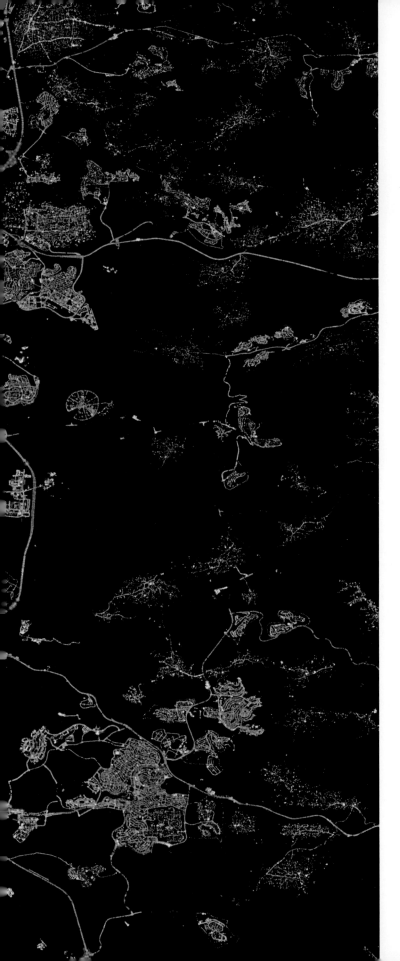

以色列 特拉维夫
Tel Aviv, Israel

特拉维夫位于以色列地中海海岸线上，是该国的经济技术中心、交通枢纽与金融产业核心。城中教育业发达，水平超越全国平均。特拉维夫绿化遍布，全年阳光充足，夏季湿热冬季温和，每年有超过 250 万名国际游客来到特拉维夫。特拉维夫的白城于 2003 年被联合国教科文组织指定为世界文化遗产，这里的建筑风格多变，包括包豪斯和其他相关的现代主义建筑风格。该城市的热门景点有雅法老城、埃雷茨以色列博物馆、艺术博物馆、哈亚尔肯公园以及海滨长廊和海滩。

Tel Aviv, is the most populous city in the Gush Dan metropolitan area of Israel. Located on the Israeli Mediterranean coastline, it is the economic, technological and financial center of the country. The education standard of the city is above the national average. Tel Aviv, which has a a lot of green spaces and a mild climate with rich sunlight, receives over 2.5 million international visitors annually. Tel Aviv's White City, designated a UNESCO World Heritage Site in 2003, comprises the world's largest concentration of International Style buildings, including Bauhaus and other related modernist architectural styles. Popular attractions include Jaffa Old City, the Eretz Israel Museum, the Museum of Art, Hayarkon Park, and the city's promenade and beach.

耶路撒冷

Jerusalem

耶路撒冷是位于西亚的一座城市，屹立在地中海和死海之间的犹太山脉高原上，全年气候温热。作为世界上最古老的城市之一，它是多个宗教的圣城，建造了许多清真寺以及圣墓教堂。由于耶路撒冷的特殊宗教历史地位，城东的老城区中工业开发少，保留了大量历史建筑和重要宗教遗址，其中包括圣殿山及其西墙、圆顶清真寺、阿克萨清真寺。新城区位于耶路撒冷西部，比较现代化，有较多的大型工商业。耶路撒冷属于地中海气候，平均高温为 12℃，年平均降水量接近 590 mm，夏季降水量低。

Jerusalem is a city in Western Asia. Situated on a plateau in the Judaean Mountains between the Mediterranean and the Dead Sea, it is one of the oldest cities in the world and is considered to be a holy city for various regions. The Old City is home to many sites of seminal religious importance, among them the Temple Mount with its Western Wall, Dome of the Rock and al-Aqsa Mosque, and the Church of the Holy Sepulchre. Western parts of the city have relatively more modern buildings and industries. Jerusalem has a Mediterranean climate with an average high temperature of 12℃ and an average annual precipitation of nearly 590 mm, with very little precipitation in the summer months from May to September.

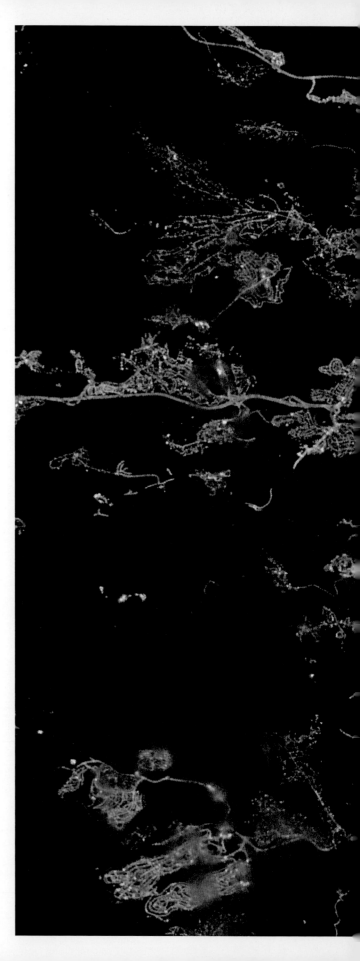

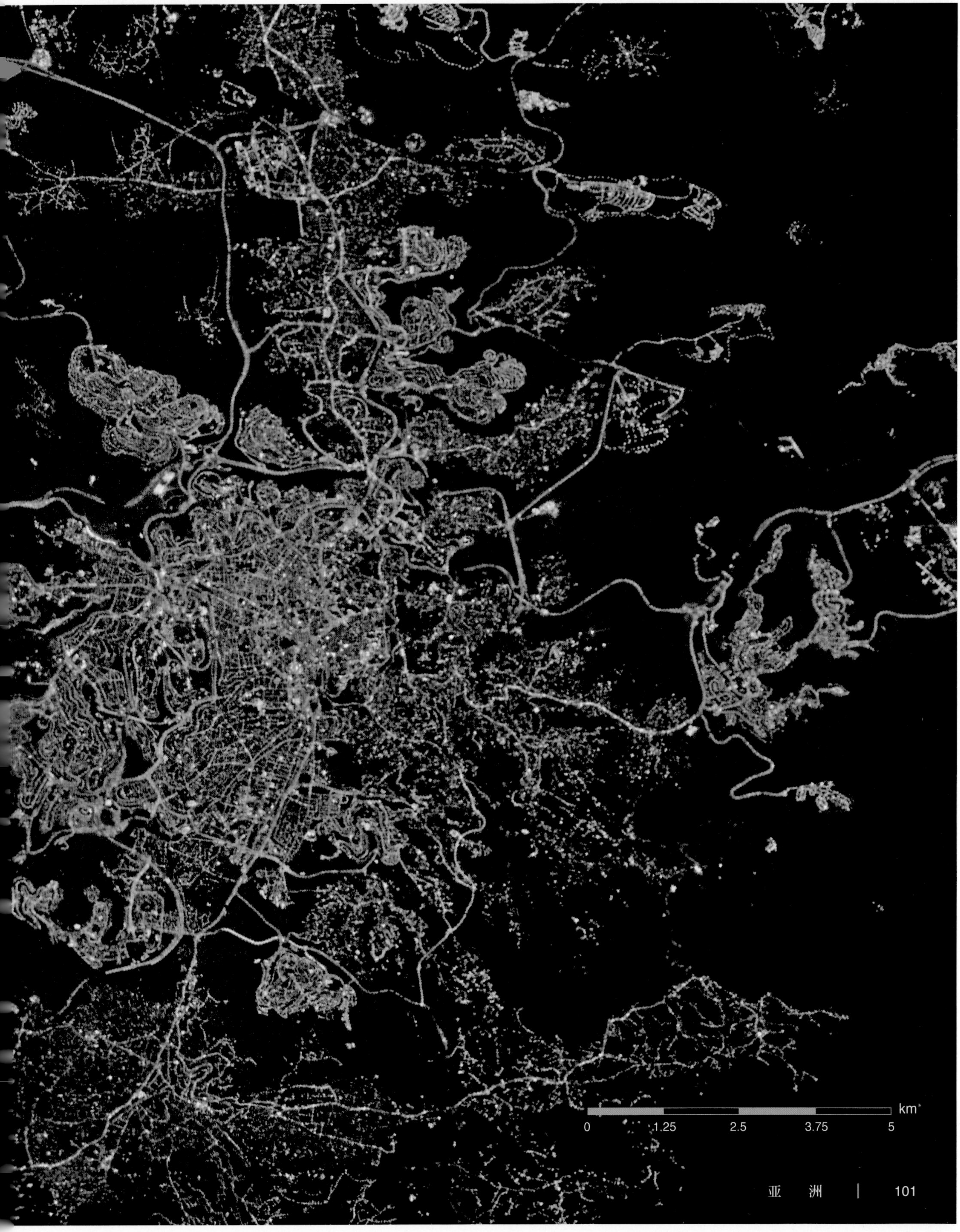
km
0
1.25
2.5
3.75
5

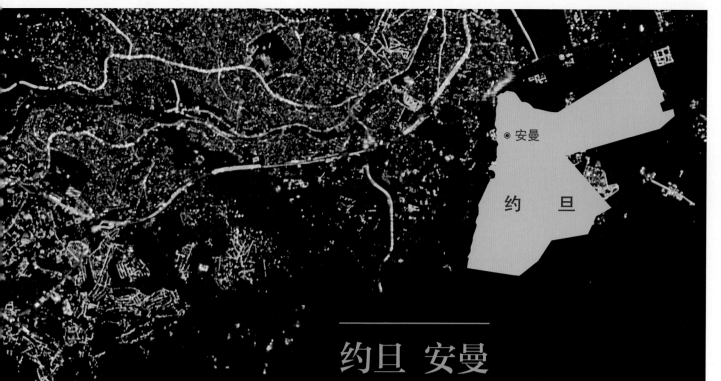

约旦 安曼
Amman, Jordan

安曼是约旦的首都和最大的城市，也是该国的经济、政治和文化中心。作为约旦的主要城市，安曼也是黎凡特地区最大的城市，是阿拉伯世界第五大城市和中东第九大都市区。城市的东部蕴藏着深厚的历史和文化，经常举办各种文化活动，而西部地区则展现出一片现代的景象，是城市的经济中心。此外，由于其良好的经济、劳动力、环境和社会文化因素，安曼被评为中东和北非地区最优秀的城市之一。安曼位于两种气候之间：西部和北部居民区为地中海气候，东部和南部居民区为半干旱气候。

Amman is the capital and largest city of Jordan, and the country's economic, political, and cultural center. Amman is Jordan's primate city and is the largest city in the Levant region, the fifth-largest city in the Arab world, and the ninth largest metropolitan area in the Middle East. East Amman is predominantly filled with historic sites that frequently host cultural activities, while West Amman is more modern and serves as the economic center of the city. Moreover, it was named one of the Middle East and North Africa's best cities according to economic, labor, environmental, and socio-cultural factors. Amman is located between two climates: a Mediterranean climate in its western and northern neighborhoods, and a semi-arid climate in its eastern and southern neighborhoods.

利雅得 ◉

沙特阿拉伯

沙特阿拉伯 利雅得
Riyadh, Saudi Arabia

利雅得位于纳吉德高原东部的纳福德沙漠中心，不仅是沙特阿拉伯的首都和最大城市，同时也是利雅得省的首府和利雅得省的中心以及阿拉伯半岛上最大的城市。这座城市的平均海拔 600 m，每年接待约 500 万游客，位列全球第 49 个最受游客欢迎的城市。作为沙特阿拉伯的政治和行政中心，利雅得汇聚了多个关键机构，包括协商会议（舒拉理事会）、部长会议、国王和最高司法委员会等。除了这四个构成沙特阿拉伯法律体系核心的机构外，其他主要和次要政府机构的总部也位于利雅得。该城市拥有 114 个外国大使馆，其中大部分位于城市西部的外交区。利雅得不仅是世界上人口增长最快的城市之一，也是许多外籍人士的家园。鉴于其政治、文化和经济重要性，利雅得在沙特阿拉伯和中东地区具有显著的地位。

Riyadh is the capital and largest city of Saudi Arabia. It is also the capital of the Riyadh Province and the center of the Riyadh Governorate. It is the largest city on the Arabian Peninsula, and is situated in the center of the an-Nafud desert, on the eastern part of the Najd plateau. The city sits at an average of 600 m above sea level, and receives around 5 million tourists each year, making it the forty-ninth most visited city in the world and the 6th in the Middle East. Riyadh is the political and administrative center of Saudi Arabia. The Consultative Assembly (also known as the Shura Council), the Council of Ministers, the King and the Supreme Judicial Council are all situated in the city. Alongside these four bodies that form the core of the legal system of Saudi Arabia, the headquarters of other major and minor governmental bodies are also located in Riyadh. The city hosts 114 foreign embassies, most of which are located in the Diplomatic Quarter in the western reaches of the city. Riyadh is one of the world's fastest-growing cities in population and is home to many expatriates.

科威特

科威特城

科威特 科威特城

Kuwait City, Kuwait

科威特城位于波斯湾科威特湾南岸的国家中心，是科威特的首都和最大的城市。科威特城是地球上夏季最炎热的城市之一，一年中有长达三个月的时间平均温度超过 45℃。科威特城是该国的政治、文化和经济中心，聚集了包括科威特的赛义夫宫、政府办公室以及大多数科威特公司和银行的总部。科威特城的贸易和运输需求主要由科威特国际机场、舒韦赫港和艾哈迈迪港等提供。

Kuwait City is the capital and largest city of Kuwait. Located at the heart of the country on the south shore of Kuwait Bay on the Persian Gulf, it is the political, cultural and economical center of the emirate, containing Kuwait's Seif Palace, government offices, and the headquarters of most Kuwaiti corporations and banks. It is one of the hottest cities in summer on earth, with average summer high temperatures over 45℃ for three months of the year. Kuwait City's trade and transportation needs are served by Kuwait International Airport, Shuwaikh Port and Port of Mena Al Ahmadi.

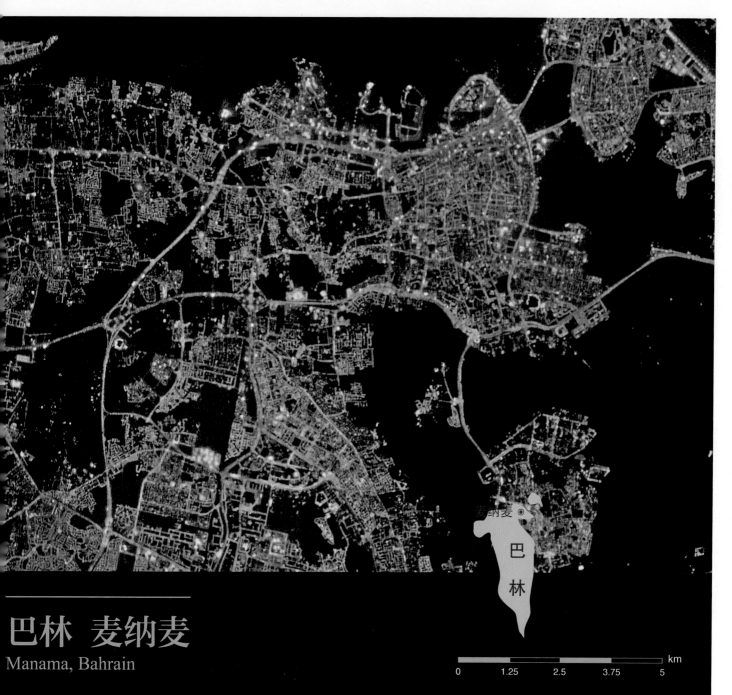

麦纳麦

巴林

km
0   1.25   2.5   3.75   5

巴林 麦纳麦
Manama, Bahrain

　　麦纳麦是巴林的首都和最大城市，是波斯湾的贸易中心之一。麦纳麦是巴林的商业之都，也是通往巴林岛的门户。20世纪的石油财富促进了该城市经济的快速增长，在20世纪90年代，巴林实施了协调一致的多元化政策推动了其他行业的扩张，帮助麦纳麦转变为中东重要的金融中心。麦纳麦被曾阿拉伯联盟指定为阿拉伯文化之都，展示了该城市在文化和艺术方面的重要性。作为一个蓬勃发展的城市，麦纳麦在中东地区发挥着重要的经济和文化作用。

Manama is the capital and largest city of Bahrain, and one of trading center in the Persian Gulf, Manama is home to a very diverse population. Manama became the mercantile capital and was the gateway to the main Bahrain Island. In the 20th century, Bahrain's oil wealth helped spur fast growth and in the 1990s a concerted diversification effort led to expansion in other industries and helped transform Manama into an important financial hub in the Middle East. Manama was designated as the 2012 capital of Arab culture by the Arab League.

卡塔尔 多哈
Doha, Qatar

　　多哈位于卡塔尔东部的波斯湾沿岸,是卡塔尔的首都和主要金融中心,也是卡塔尔发展最快的城市和商业中心。20 世纪初,多哈从一个拥有约 350 艘采珠船的渔村转型为一个现代化、繁荣的城市。该市拥有许多主要景点和博物馆,成为人们关注的焦点。多哈以其热情好客、精致的美食、独特的购物体验、传统的露天集市以及丰富的探险活动而闻名,为所有旅行者提供了丰富的体验。多哈曾举办第 20 届世界石油大会、2012 年《联合国气候变化框架公约》气候谈判以及 2022 年的足球世界杯,进一步推动了多哈的国际知名度和地位。

　　Doha is the capital city and main financial hub of Qatar. Located on the Persian Gulf coast in the east of the country, it is the country's fastest-growing city and its business center. From a pearling and fishing village of great importance, with around 350 pearling boats at the beginning of the 20th century, Doha underwent a complete economic transformation and became a modern and prosperous city. Doha's tourism sector has evolved rapidly over the years, with key attractions and several museums of great importance gaining the spotlight. Known for its exemplary hospitality, exquisite cuisine, unique shopping experiences and traditional souqs, along with a plethora of adventures and activities, Doha is growing rapidly, and offers something for all travelers. The World Petroleum Council held the 20th World Petroleum Conference in Doha. Additionally, the city hosted the 2012 UNFCCC Climate Negotiations and the 2022 FIFA World Cup.

多哈 ◎

卡塔尔

km

0　　2　　4　　6　　8

也 门

萨那

也门 萨那

Sana'a , Yemen

　　萨那是也门的首都和最大的城市，同时也是萨那省的行政中心。萨那海拔 2300 ㎡，是世界上海拔最高的首都之一；其位置不仅是该国最高的山脉，也是该地区最高的山脉之一。萨那属于沙漠气候，但气候温和，年平均气温 17.5℃，年降水量 300 mm。萨那的旧城被联合国教科文组织列为世界遗产，它的建筑特点独具一格，最引人注目的是其装饰有几何图案的多层建筑。在旧城内，萨利赫大清真寺尤为显眼，作为萨那最大的清真寺，它见证了这座古城的历史。

　　Sana'a is the capital and largest city in Yemen and the center of Sana'a Governorate. At an elevation of 2300 m, Sana'a is one of the highest capital cities in the world, considered to be the highest mountains in the country and amongst the highest in the region. Sana'a has a desert climate but a mild one, with an average annual temperature of 17.5℃ and an annual precipitation of 300 mm. The Old City of Sana'a, a UNESCO World Heritage Site, has a distinctive architectural character, most notably expressed in its multi-storey buildings decorated with geometric patterns. The Al Saleh Mosque, the largest in Sana'a, is located in the Old City.

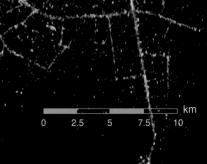

阿拉伯联合酋长国
阿布扎比
Abu Dhabi, The United Arab Emirates

阿布扎比是阿拉伯联合酋长国的首都和人口第二大城市。它同时也是阿布扎比酋长国的首府和阿布扎比大都会区的中心。阿布扎比市位于波斯湾的一个岛屿上，远离阿联酋中西部海岸。得益于其迅速的城市发展、丰富的石油和天然气资源，以及相对较高的平均收入，阿布扎比已发展成为一个大型、发达的大都市，其在阿联酋和整个中东地区具有重要的政治、经济、文化和商业地位。

Abu Dhabi is the capital and the second-largest city of the United Arab Emirates.It also serves as the capital of the Emirate of Abu Dhabi and as the center of the Abu Dhabi Metropolitan Area. Situated on an island in the Persian Gulf, Abu Dhabi is distanced from the central and western coasts of the UAE. Thanks to its rapid urban development, abundant oil and natural gas resources, and relatively high average income, Abu Dhabi has evolved into a large and prosperous metropolis. It holds significant political, economic, cultural, and commercial importance within the UAE and across the broader Middle East region.

阿布扎比

阿拉伯联合酋长国

km
0   1   2   3   4

阿拉伯联合酋长国　迪拜
Dubai, United Arab Emirates

迪拜是阿拉伯联合酋长国人口最多的城市，同时也是其中 7 个酋长国中人口最多的迪拜酋长国的首都。起初，这座城市在 18 世纪只是一个小渔村，但在 21 世纪初迅速发展，尤其是在旅游业和奢侈品领域。迪拜引以为豪的特点之一是拥有世界上第二多的五星级酒店，以及世界上最高的建筑——828 m 高的哈利法塔。迪拜位于波斯湾沿岸的阿拉伯半岛东部，是全球主要的客运和货运枢纽。得益于石油产业带来的财富，迪拜迅速崛起为一个重要的国际商业枢纽。自 20 世纪初以来，迪拜一直是地区和国际贸易的中心，其经济依赖于贸易、旅游、航空、房地产和金融服务等领域的收入。迪拜以其现代化的城市规划、豪华的购物中心、壮观的建筑和独特的文化体验吸引了来自世界各地的人们，形成了多样化和充满活力的社会，是世界上备受瞩目的旅游和商务目的地之一。

Dubai is the most populous city in the United Arab Emirates and the capital of the Emirate of Dubai, the most populous of the seven emirates. Initially a small fishing village in the 18th century, the city grew rapidly at the beginning of the 21st century, especially in the field of tourism and luxury goods. One of Dubai's proud features is the second largest number of five-star hotels in the world, as well as the world's tallest building, the 828-metre-high Burj Khalifa. Located on the eastern part of the Arabian Peninsula along the Persian Gulf, Dubai has become a major global hub for passenger and cargo transport. Thanks to the wealth generated by the oil industry, Dubai has rapidly emerged as a major international business hub. Dubai has been a centre of regional and international trade since the early 20th century, and its economy depends on revenues from trade, tourism, aviation, real estate and financial services. Dubai attracts people from all over the world with its modern urban planning, luxury shopping malls, spectacular architecture and unique cultural experiences, resulting in a diverse and vibrant society that is one of the world's most highly regarded tourist and business destinations.

迪拜

阿拉伯联合酋长国

马斯喀特 ◉

阿　曼

阿曼 马斯喀特
Muscat, Oman

　　马斯喀特是阿曼的首都，同时也是阿曼人口最多的城市。作为阿曼湾的一个重要港口城市，马斯喀特吸引了波斯人、俾路支人和信德人等外国商人和定居者。这座城市位于阿曼湾沿岸的阿拉伯海岸，靠近具有战略意义的霍尔木兹海峡。马斯喀特城区的大多数建筑为低矮的白色建筑，穆特拉港区则位于城市的东北边缘，拥有海堤和港口。马斯喀特气候干旱，夏季漫长闷热，冬季温暖。

Muscat is the capital and most populated city in Oman. As an important port-town in the Gulf of Oman, Muscat attracted foreign tradesmen and settlers such as the Persians, Balochis and Sindhis. The city lies on the Arabian Sea along the Gulf of Oman and is in the proximity of the strategic Straits of Hormuz. Low-lying white buildings typify most of Muscat's urban landscape, while the port-district of Muttrah, with its corniche and harbor, form the north-eastern periphery of the city. Muscat has arid climate with long, sweltering summers and warm winters.

km

0   1.25   2.5   3.75   5

印度 新德里
New Delhi, India

　　新德里位于印度的西北部，喜马拉雅山脉西方。恒河支流亚穆纳河从城东缓缓流过，河对岸是广阔的恒河平原。新德里是在古老的德里城基础上扩建而成。新德里和老德里中间隔着一座印度门，印度门以南为新德里，印度门以北为老德里。新德里以姆拉斯广场为中心，城市街道呈辐射状、蛛网式地伸向四面八方，市中心建造有较多宏伟的建筑。主要政府机构集中在从印度总统府到印度门之间绵延几公里的宽阔大道两旁。新德里属亚热带季风气候，春夏之交较热，每年雨季从6月中旬开始，持续到10月初。

New Delhi is located in the north-west of India, near the southern part of the western Himalayas. The Yamuna River, a tributary of the Ganges, flows gently from the east of the city, and on the opposite bank of the river is the vast Ganges Plain. New Delhi was built on the site of the old city of Delhi. New Delhi and Old Delhi are separated by the Indian Gate, with New Delhi to the south of the Indian Gate and Old Delhi to the north of the Indian Gate. New Delhi is centred on Mallas Square, and the city's streets stretch out in all directions in a radial, web-like pattern. Most of the significant architectural complexes are concentrated in the city center. The main government institutions are concentrated in the city along the wide avenues that stretch for several kilometers from the President of India's residence to India Gate. New Delhi has a subtropical monsoon climate with hotter spring and summer months. The annual rainy season begins on average in mid-June and lasts until early October.

乌兰巴托 ◉

蒙　古

```
                                                    km
0        1.25       2.5       3.75        5
```

蒙古　乌兰巴托

Ulaanbaatar, Mongolia

　　乌兰巴托是蒙古的首都和人口最多的城市。该市位于蒙古中北部，海拔约 1300 m，坐落在土兀剌河流域的一个山谷中。乌兰巴托建于 1639 年，最初是一个游牧佛教寺院的中心，经历了 28 次迁移，直至 1778 年定址。1924 年，随着蒙古人民共和国的成立，这座城市正式更名为乌兰巴托，并宣布为蒙古的首都。现代城市规划始于 20 世纪 50 年代，大部分传统的蒙古包被苏联风格的公寓所取代。乌兰巴托是蒙古的文化、工业和金融中心，也是蒙古交通网络的重要枢纽，通过铁路与俄罗斯的西伯利亚大铁路和中国的铁路系统相连。

Ulaanbaatar is the capital and most populous city of Mongolia. The municipality is located in north central Mongolia at an elevation of about 1300 m in a valley on the Tuul River. The city was originally founded in 1639 as a nomadic Buddhist monastic center, changing location 28 times, and was permanently settled at its current location in 1778. With the proclamation of the Mongolian People's Republic in 1924, the city was officially renamed Ulaanbaatar and declared the country's capital. Modern urban planning began in the 1950s, with most of the old Ger districts replaced by Soviet-style flats. It contains almost half of the country's total population. As the country's primate city, it serves as the cultural, industrial and financial heart as well as the center of Mongolia's transport network, connected by rail to both the Trans-Siberian Railway in Russia and the Chinese railway system.

朝鲜 平壤

Pyongyang, North Korea

平壤位于朝鲜半岛西北部，大同江横跨其中，因其地势平坦而得名。大同江与其支流流经市中心。平壤是朝鲜半岛历史最悠久的城市，保留着高句丽古城、安鹤宫遗址、广法寺等历史古迹，该城市面积 2629.4 km²，下设 18 个区、4 个郡，年平均气温 9.7℃

Pyongyang is located in the northwest of the Korean Peninsula, with the Taedong River running through it. It is named for its flat terrain. The Taedong River and its tributaries flow through the city center. Pyongyang is the oldest city on the Korean Peninsula and preserves historical sites such as the ancient city of Goguryeo, the site of Anhe Palace, and Gwanghwamun Temple. The city covers an area of 2629.4 km² and is divided into 18 districts and 4 counties. The average annual temperature is 9.7℃.

朝鲜

◎ 平壤

km
0 1 2 3 4

韩国 首尔

Seoul, South Korea

　　首尔被群山环绕，是韩国的首都和第一大都市。首都圈以江南和数码媒体城为中心，内有昌德宫、朝鲜王陵等 5 处世界文化遗产，同时也是三星、LG、现代等 15 家世界 500 强企业的总部所在地，曾主办过 1986 年亚运会和 1988 年夏季奥运会。首尔位于汉江沿岸的战略位置，是历史上多个政权的首都。首尔还是韩国流行音乐和韩流的发源地，也是韩国文化和大型媒体、娱乐公司和广播公司的发源地，曾被评为世界设计之都。

　　Seoul, the capital and largest metropolis of South Korea, is surrounded by mountains. With major technology hubs centered in Gangnam and Digital Media City, the Seoul Capital Area is home to the headquarters of five UNESCO World Heritage Sites containing Changdeok Palace, the Royal Tombs of the Joseon Dynasty and so on, as well as 15 Fortune Global 500 companies, including Samsung, LG, Hyundai and so on. Seoul has hosted the 1986 Asian Games, and the 1988 Summer Olympics. Strategically located along the Han River, Seoul was the capital of several regimes in history. Seoul was named the World Design Capital, as it is the birthplace of Korean popular cultures and the heart of mass medias, entertainment firms, and broadcasters.

km

0　　　2.5　　　5　　　7.5　　　10

首尔

韩 国

不丹 廷布
Thimphu, Bhutan

　　廷布是不丹的首都，位于不丹的中西部。廷布是世界海拔第五高的首都，海拔从 2248 m 到 2648 m 不等，是不丹的政治和经济中心，主要产业是农业和畜牧业。廷布还是不丹文化和宗教的中心。城市内有许多佛教寺庙、修道院和宗教遗址，吸引着朝圣者和游客。同时，廷布也是不丹文化的传承地，经常举办各种节日、庆典和文化活动。

　　Thimphu is the capital city of Bhutan. It is situated in the western central part of Bhutan. Thimphu is the fifth highest capital in the world by altitude and ranges in altitude from 2248 m to 2648 m. Thimphu, as the political and economic center of Bhutan, has a dominant agriculture and livestock base. Thimphu is an essential hub for Bhutanese culture and religion. The city boasts numerous Buddhist temples, monasteries, and religious sites that allure pilgrims and tourists alike. Concurrently, Thimphu serves as a cradle for Bhutanese cultural heritage, hosting a plethora of festivals, celebrations, and cultural events.

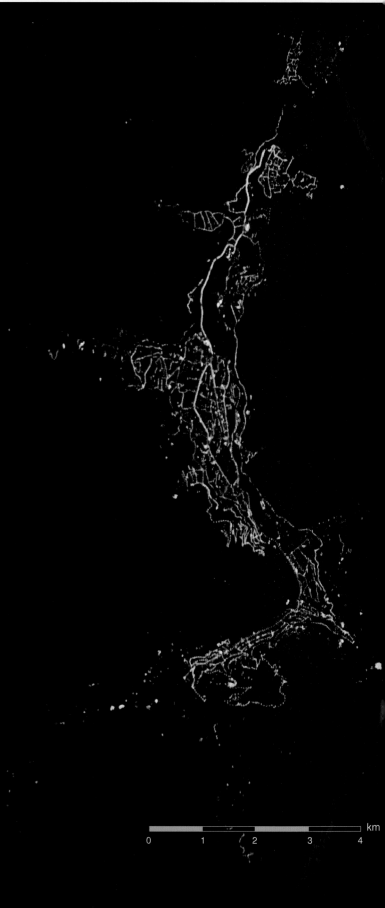

km

0　　　1　　　2　　　3　　　4

孟加拉国 达卡
Dhaka, Bangladesh

　　达卡作为孟加拉国的首都和国家政治、经济和文化生活中心，是孟加拉国许多公司以及教育、科学、研究和文化组织的所在地。自成为现代化的首都以来，达卡的人口、面积以及社会和经济多样性有巨大提升。这座城市现在是孟加拉国工业化最密集的地区之一，该市的经济贡献占孟加拉国经济的三分之一，其中达卡证券交易所拥有 750 多家上市公司。达卡以人力车、比里亚尼（传统的三轮车）、艺术节和宗教多样性闻名。

As the capital city of Bangladesh and the center of political, economic and culture life in Bangladesh, Dhaka hosts many Bangladeshi companies and leading Bangladeshi educational, scientific, research and cultural organizations. Since its establishment as a modern capital city, the population, area and social and economic diversity of Dhaka have grown tremendously. The city is now one of the most densely industrialized regions in the country. The city accounts for one-third of Bangladesh's economy. The Dhaka Stock Exchange has over 750 listed companies. The city is known for its rickshaws, biryani, art festivals and religious diversity.

孟加拉

达卡

km
0　1.25　2.5　3.75　5

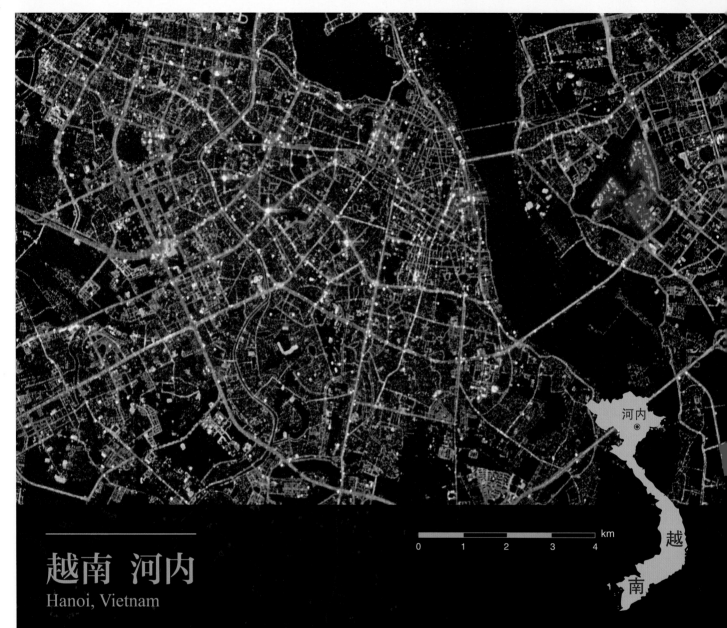

越南 河内
Hanoi, Vietnam

　　河内是越南的首都和第二大城市,位于越南北部的红河三角洲。河内的历史可以追溯到公元前三世纪。作为越南的主要旅游目的地,河内拥有许多保存完好的 20 世纪的建筑以及佛教、天主教、儒教和道教的宗教遗址,还有越南帝国时期的几个历史地标和大量博物馆。其中,昇龙皇城的中央部分于 2010 年被联合国教科文组织认定为世界遗产,标志着其重要历史价值得到国际认可。河内作为一个充满活力和魅力的城市,持续吸引着来自世界各地的游客。

Hanoi is the capital and second-largest city of Vietnam, located within the Red River Delta of Northern Vietnam. Hanoi can trace its history back to the third century BC. As a major tourist destination in Vietnam, Hanoi offers well-preserved French colonial architecture, religious sites dedicated to Buddhism, Catholicism, Confucianism and Taoism, several historical landmarks of Vietnamese imperial periods, and a large number of museums. The central part of Thang Long Imperial Citadel was recognized as a World Heritage Site by UNESCO in 2010, which marks its important historical value has been recognized internationally. As a vibrant and charming city, Hanoi continues to attract tourists from all over the world.

缅甸 内比都
Nay Pyi Taw, Myanmar

缅甸

内比都

    内比都是缅甸的首都，也是缅甸境内的第三大城市，该城市面积大，而人口密度极低。这座城市位于缅甸中部山区的彬马那，不仅是缅甸首都政府的所在地，也是联邦议会、最高法院、总统府、缅甸内阁官邸以及政府各部和军队总部所在地。

    Nay Pyi Taw is the capital of Myanmar and the third largest city in Myanmar, known for its large area and extremely low population density. The city is located in in Binmana, a mountainous area in central Myanmar, with an important strategic position, which is not only the home to the capital government of Myanmar, but also the seat of the Federal Parliament, the Supreme Court, the Presidential Palace, the official residence of the Myanmar Cabinet, and the headquarters of various government ministries and the army.

km

0   1   2   3   4

老挝 万象
Vientiane, Laos

　　万象是老挝的首都和最大的城市，也是老挝的经济中心。万象位于湄公河岸边，靠近泰国边境。该城市还坐落着老挝重要的国家级历史遗迹——塔銮，是老挝的标志性建筑，也是老挝的佛教圣地。城中佛寺林立，其中著名寺庙包括玉佛寺。

　　Vientiane, the capital and largest city of Laos, is the economic center of Laos. It is located on the banks of the Mekong River, close to the Thailand border. Pha That Luang, the most significant national monuments of Laos, is located in Vientiane, which is a known symbol of Laos and an icon of Buddhism in Laos. Other significant Buddhist temples in Laos can be found in the city as well, such as Haw Phra Kaew.

老挝

万象

挝

km

| 0 | 1.25 | 2.5 | 3.75 | 5 |

菲律宾 马尼拉

Manila, the Philippines

　　马尼拉位于吕宋岛马尼拉湾东岸，是菲律宾的首都。帕西格河流经城市中部，将城市分为南北两部分。马尼拉深度参与全球贸易，在历史上，通过该城的帆船商业网络，开创了跨太平洋，连接亚洲和拉美国家的贸易。马尼拉全境位于热带，四季如夏。由于其纬度接近赤道，气温的变化范围较小，全年平均气温为 25 ～ 31℃。

Manila is located on the east coast of Manila Bay on Luzon Island and is the capital of the Philippines. The Pasig River flows through the center of the city, dividing the city into two parts, north and south. Manila is deeply involved in global trade, historically pioneering the trans-Pacific trade linking Asia and Spanish Americas through the city's merchant network of galleons. The whole territory of Manila is located in the tropics, and it is like summer all year round. Because its latitude is close to the equator, the temperature range is small, and the annual average temperature is 25℃ to 31℃ .

马尼拉

菲律宾

km

0   1.25   2.5   3.75   5

泰国 曼谷
Bangkok, Thailand

　　曼谷是泰国的首都和人口最多的城市，被誉为"天使之城"，位于泰国中部的湄南河三角洲，占地 1568.7 km²。曼谷的历史可以追溯到 15 世纪，原来是一个渔村，在吞武里王朝兴起时，逐渐形成了一些小集市和居民点，随后发展壮大，并最终成为首都。如今，曼谷已成为区域性的金融和商业中心，也是国际交通枢纽和医疗保健的枢纽，同时成为了艺术、时尚和娱乐中心。这座城市以其繁华的街头生活、丰富的文化地标以及寺庙闻名，其中包括广为人知的大皇宫、郑王庙和卧佛寺等佛教寺庙。作为世界级的旅游目的地之一，曼谷吸引全球各地的游客，多年来在多项国际排名中一直被评为世界上最受欢迎的城市。

Bangkok is the capital and most populous city of Thailand, known as the City of Angels, located in the Chao Phraya River Delta in central Thailand, covering an area of 1568.7 km². The history of Bangkok can be traced back to the 15th century. It was originally a fishing village. When the Thonburi Dynasty rose, some small markets and settlements gradually formed, and then developed and expanded, and eventually became the capital. The city is now a regional force in finance and business. It is an international hub for transport and health care, and has emerged as a center for the arts, fashion, and entertainment. The city is known for its prosperous street life and cultural landmarks. The Grand Palace and Buddhist temples including Wat Arun and Wat Pho stand in contrast with other tourist attractions. Bangkok is among the world's top tourist destinations, and has been named the world's most visited city consistently in several international rankings.

柬埔寨 金边

Phnom Penh, Cambodia

　　金边始建于 1372 年，以其历史建筑和景点而闻名。它位于洞里萨河、湄公河和巴萨河两岸，人口超过 200 万，约占柬埔寨人口的 14%，是柬埔寨人口最多的城市。金边长久以来一直是柬埔寨的首都，并已发展成为该国的主要城市及其经济、工业和文化中心。

Phnom Penh was founded in 1372 and is famous for its historical buildings and attractions. It is located on the banks of the Tonle Sap River, the Mekong River and the Basa River, with a population of more than 2 million, accounting for about 14% of the Cambodian population, making it the most populous city in Cambodia. Phnom Penh has long been the capital of Cambodia and has grown into the country's main city and its economic, industrial and cultural center.

柬埔寨

金边 ◎

km

0   1.25   2.5   3.75   5

马来西亚

马来西亚 吉隆坡
Kuala Lumpur, Malaysia

　　吉隆坡是马来西亚的首都，占地面积为 243 km²。它是亚洲发展最快的城市之一，也是马来西亚的文化、金融和经济中心。吉隆坡起初只是锡矿产业的城镇，经历了近几十年的快速发展，成为了一个现代化的国际都市。它拥有世界上最高的双子塔——吉隆坡石油双塔，是见证马来西亚发展的标志性建筑。吉隆坡以其城市风貌和活力而闻名，是世界著名的旅游购物城市，它拥有多个世界级购物中心，其中有三个被列入世界十大购物中心。

　　Kuala Lumpur is the capital of Malaysia and covers an area of 243 km². It is one of the fastest growing cities in Asia and the cultural, financial and economic center of Malaysia. Kuala Lumpur started as a town serving the tin mines in the region, and has undergone rapid development in recent decades to become a modern international city. It houses the world's tallest twin buildings, the Petronas Twin Towers, which have since become an iconic symbol of Malaysia's development. Kuala Lumpur, known for its urban style and vitality, is a world-famous tourist shopping city. It has many world-class shopping malls, three of which are listed in the world's top ten shopping malls.

km
0   1.25   2.5   3.75   5

新加坡
Singapore

　　新加坡位于赤道以北约 137 km，向北与马来半岛南端隔海相望，西临马六甲海峡，南临新加坡海峡，东临中国南海。该国领土由一个主岛、63 个卫星岛屿和小岛组成。自国家独立以来，由于大规模的填海，国土总面积增加了 25%。新加坡是世界上人口密度第三高的国家，拥有多元文化的人口，设有四种官方语言：英语、马来语、普通话和泰米尔语。新加坡在教育、医疗、基建和住房等关键社会指标上排名很高，是世界上预期寿命最长、上网速度最快、婴儿死亡率最低的国家之一。这个国家以其高效的城市规划、现代化的基础设施和独特的文化多样性而著称，是一个充满活力的繁荣地区。

　　Singapore is located about 137 km north of the equator, across the sea from the southern tip of the Malay Peninsula to the north, the Strait of Malacca to the west, the Strait of Singapore to the south, and the South China Sea to the east. The country's territory is composed of one main island, 63 satellite islands and islets. Since the country's independence, the total area of these areas has increased by 25% due to massive land reclamation. With the third highest population density in the world, Singapore has a multicultural population. Singapore has four official languages: English, Malay, Mandarin, and Tamil. Ranking highly on key social indicators such as education, healthcare, infrastructure and housing, Singapore has one of the world's longest life expectancies, fastest internet speeds and lowest infant mortality rates. Known for its efficient urban planning, modern infrastructure and unique cultural diversity, the country is a vibrant and thriving region.

新加坡

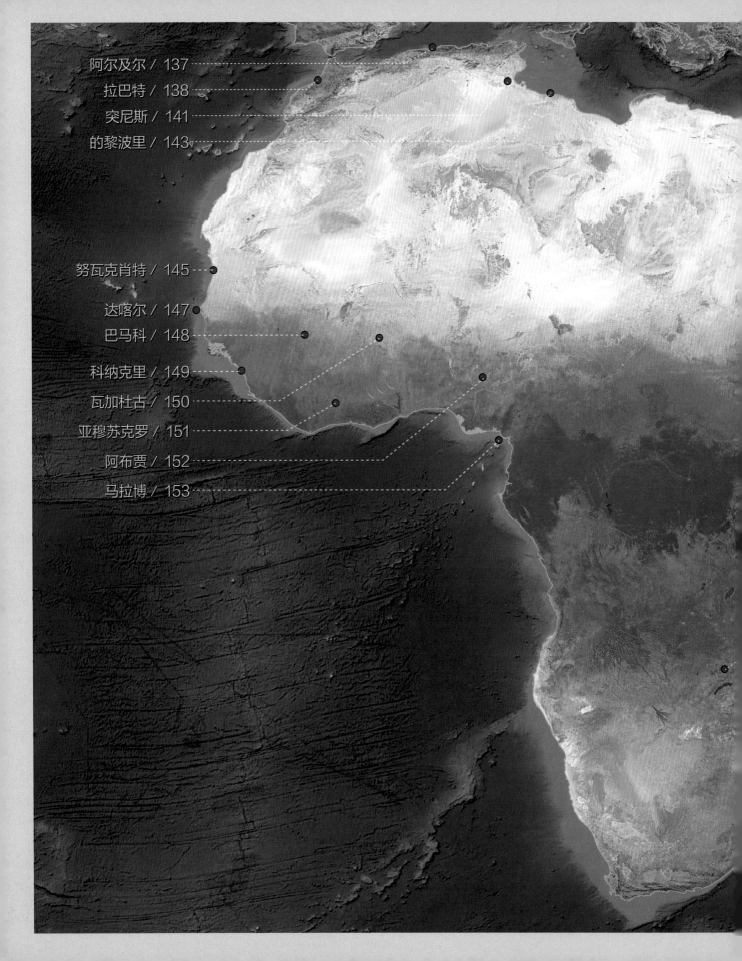

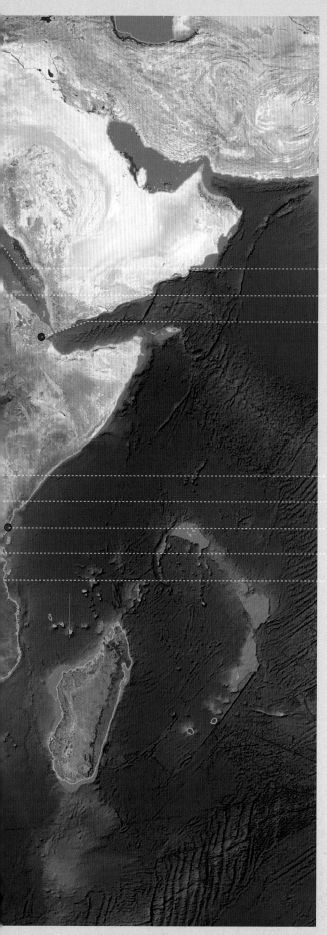

非洲

AFRICA

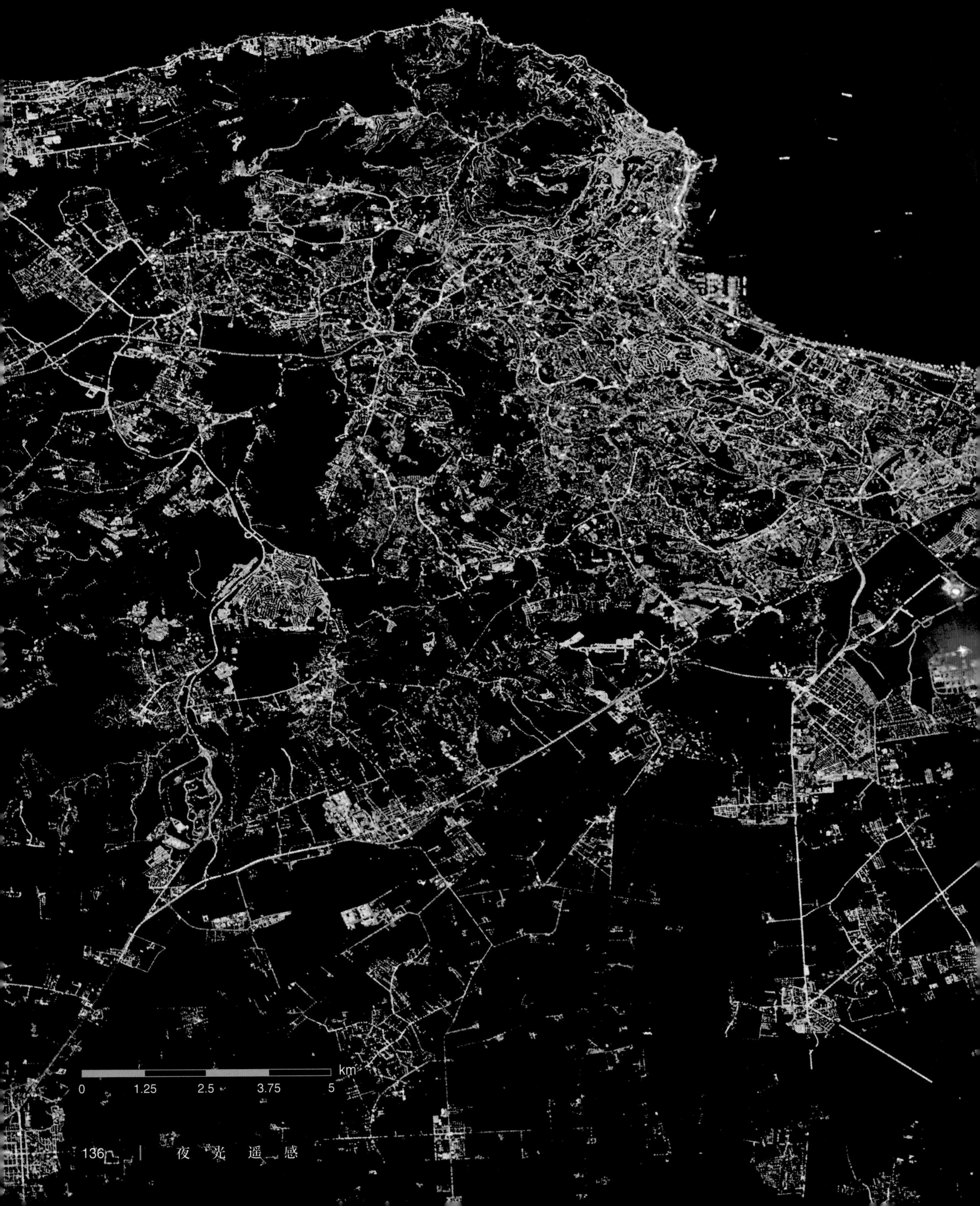
km
0
1.25
2.5
3.75
5

阿尔及利亚 阿尔及尔
Algiers, Algeria

阿尔及尔是阿尔及利亚的首都和最大的城市，位于地中海西岸、阿尔及利亚的中北部。以其优美的白色建筑群和独特的奥斯曼式建筑风格赢得了"白色阿尔及尔"的美誉。阿尔及尔是阿尔及利亚重要的经济、商业和金融中心，在石油和天然气产业中具有重要地位；同时阿尔及尔也是阿尔及利亚的文化和学术中心，拥有阿尔及利亚国家图书馆、贝尔索阿美术馆、阿尔及尔国家公园等众多文化和历史景点，以及阿尔及尔大学和其他高等教育机构。阿尔及尔在阿拉伯世界中的地位举足轻重，举办了众多区域性和国际性活动，如阿拉伯联盟峰会和阿尔及尔国际电影节。阿尔及尔以其充满活力的音乐和艺术场景，尤其是 20 世纪初起源于阿尔及尔的拉伊音乐，使其在世界文化舞台上独具一格。

Algiers, the capital and largest city of Algeria, sits on the western coast of the Mediterranean in the north-central part of the country. Algiers is now known as the White Algiers due to its beautiful white architecture and distinctive Ottoman style. Algiers serves not only as Algeria's main economic, commercial, and financial center, particularly pivotal in the oil and natural gas industries, but also its cultural and academic hub, housing numerous cultural and historical sites including the National Library of Algeria, the Belouizdad Art Museum, the Algiers National Park, and higher education institutions such as the University of Algiers. Its significance in the Arab world is profound, hosting numerous regional and international conferences and festivals like the Arab League Summit and Algiers International Film Festival. With its vibrant music and art scene, especially the Rai music originating from Algiers in the early 20th century, it holds a unique place on the world's cultural stage.

摩洛哥 拉巴特

Rabat, Morocco

　　拉巴特是摩洛哥的首都，也是其第七大城市，同时担任拉巴特－塞勒－肯尼特拉行政区首府，位于大西洋上的布雷格河河口，对岸是通勤城镇塞尔。12世纪，阿尔莫哈德王朝在此建城，如今的拉巴特繁华而活跃，老与新、传统与现代巧妙融合，是诸多外国使馆所在地，也是重要的文化中心，拥有众多的博物馆和艺术场所。城市的建筑遗产得到了保存完好，城市结构和许多公共建筑反映出西方和东方的双重影响。拉巴特是摩洛哥四大皇家城市之一，城中的麦地那被列为世界遗产，其丰富的历史、文化以及一系列传统商店、历史地标和蜿蜒的街道展示了城市的历史活力。乌达亚斯堡垒是城市的另一标志性历史遗址，可以俯瞰布雷格河及对岸的塞尔。

　　Rabat is the capital city of Morocco and the country's seventh largest city. It is also the capital city of the Rabat-Sale-Kenitra administrative region. Rabat is located on the Atlantic Ocean at the estuary of the Brag River, opposite Sale, the city's main commuter town. In the 12th century, the Almohad Dynasty built a city here. Today, Rabat is prosperous and active, combining old and new, tradition and modernity. It is the seat of many foreign embassies and an important cultural center, with rich museums and art venues. The city's architectural heritage is well preserved, with the urban fabric and many public buildings reflecting both Western and Eastern influences. Rabat is one of the four royal cities of Morocco. The Medina is listed as a UNESCO World Heritage Site. Its rich history, culture and a series of traditional shops, historical landmarks and winding streets show the historical vitality of the city. Udayas Fortress is another iconic historical site of the city, overlooking the Brag River and the opposite bank of Sale.

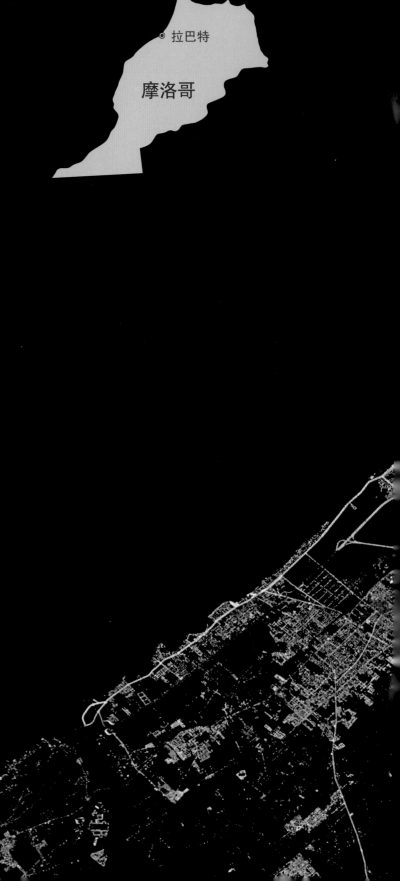

km

0　　1.25　　2.5　　3.75　　5

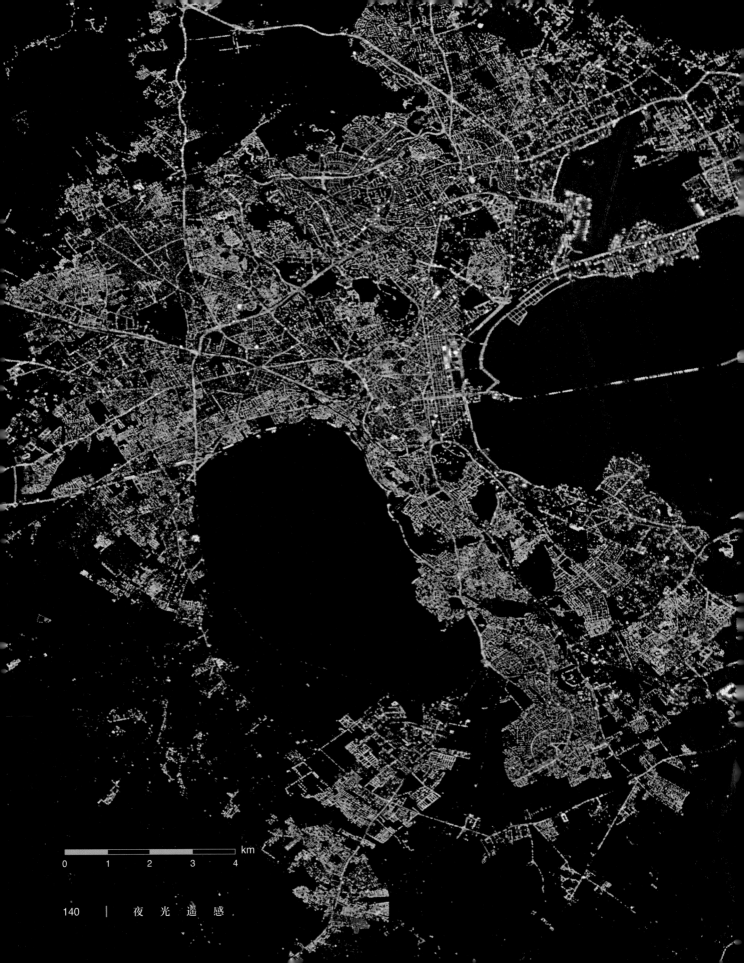

突尼斯 突尼斯
Tunis, Tunisia

突尼斯市，作为突尼斯的首都和最大城市，拥有约270万人口的大都市区。这座城市位列马格里布地区的第四大城市，也是阿拉伯世界的大城市之一。它坐落在广阔的地中海海湾——突尼斯湾的边缘，沿着突尼斯湖和拉古莱特港展开，其城市边界延伸至沿海平原和周围的山丘。城市的核心是被评为世界文化遗产的麦地那。从麦地那向东穿过海门，就进入了被称为"新城"的现代城市区域，这里以哈比布·布尔吉巴大道为主轴，沿街殖民时代的建筑与小型的古建筑形成了鲜明的对比。作为国家的首都，突尼斯市不仅是突尼斯政治和行政的中心，也是该国商业和文化活动的中心。

Tunis, as the capital and the largest city of Tunisia, boasts a metropolitan area, often referred to as "Greater Tunis", with approximately 2.7 million residents. This city ranks as the fourth largest in the Maghreb region and the one of the largest cities in the Arab world. It is situated on the edge of the vast Gulf of Tunis in the Mediterranean Sea, stretching along the Lake of Tunis and the port of La Goulette, extending its urban boundaries to the coastal plains and surrounding hills. The heart of the city is the ancient Medina, a UNESCO World Heritage Site. Eastward from the Medina, through the Sea Gate, one enters the so-called "New City", where the grand Avenue Habib Bourguiba is lined with colonial-era buildings juxtaposing smaller ancient structures. As the nation's capital, Tunis serves not only as the hub of Tunisian politics and administration, but also as the center of the country's commercial and cultural activities.

0 1.25 2.5 3.75 5

km

的黎波里

利 比 亚

利比亚 的黎波里
Tripoli, Libya

　　的黎波里是利比亚的首都，位于利比亚西北部沙漠的边缘及地中海沿岸，也是该国的最大城市、主要港口、最大商业和制造中心，拥有丰富的历史和文化遗产。的黎波里的历史可以追溯到公元前 7 世纪，历史与政权的变迁使得这座城市的建筑风格混合多元。其中，老城区的狭窄街道和传统市场都反映出了这座城市深厚的历史底蕴。在经济上，的黎波里是利比亚的经济中心，以石油、贸易和制造业为主导，同时，其丰富的海洋资源也使得渔业发展充满潜力。由于其地理位置优越，的黎波里的港口是非洲北部最大的港口之一，是利比亚与欧洲、地中海地区以及其他非洲国家的重要贸易通道。

　　Tripoli is the capital of Libya, located on the edge of the desert in the northwest of the country and along the Mediterranean Sea. It is also the largest city, main port, and largest commercial and manufacturing center in the country, boasting a rich historical and cultural heritage. The history of Tripoli dates back to the 7th century BC, and the city has been ruled by Romans, Ottomans, and Italians, resulting in a city with diverse architectural styles. The narrow streets and traditional markets of the old town reflect the deep historical roots of the city. Economically, Tripoli is the main economic center of Libya, dominated by oil, trade, and manufacturing, and its abundant marine resources also give great potential to the development of the fishing industry. Due to its advantageous geographical location, the port of Tripoli is one of the largest ports in North Africa and an important trade route for Libya with Europe, the Mediterranean region, and other African countries.

毛里塔尼亚

°努瓦克肖特

毛里塔尼亚 努瓦克肖特
Nouakchott, Mauritania

努瓦克肖特是毛里塔尼亚的首都和最大城市，坐落于撒哈拉沙漠西部，距大西洋仅 5 km，是萨赫勒地区的主要城市之一。努瓦克肖特是毛里塔尼亚的经济中心，也是毛里塔尼亚主要港口和最大的机场所在地，对外贸出口至关重要。同时，作为文化中心，该城市中设有努瓦克肖特大学及多所专业高等教育机构，并有国家博物馆及众多艺术画廊。每年冬季，大批迁徙鸟类在该城市暂留，使其成为观鸟爱好者的天堂。

Nouakchott, the capital and largest city of Mauritania, is located on the western edge of the Sahara Desert, just 5 km from the Atlantic Ocean and is one of the main cities in the Sahel region. Nouakchott is the economic center of Mauritania, as well as the location of Mauritania's main port and largest airport, which is crucial for foreign trade exports. It is also a cultural hub with Nouakchott University and several specialized higher education institutions, along with a national museum and numerous art galleries. In addition, it turns into a haven for birdwatchers every winter when a large number of migratory birds stop here.

km

0   1   2   3   4

达喀尔

塞内加尔

°达喀尔

km
0   1   2   3   4

塞内加尔 达喀尔
Dakar, Senegal

　　达喀尔坐落于大西洋沿岸的佛得角半岛上，是塞内加尔的首都和最大城市。作为非洲大陆最西端的城市，它在跨大西洋和欧洲贸易中具有至关重要的地位，是一个重要的海港。城市内拥有多所大学和研究机构，其中包括塞内加尔大学，因此成为了西非的教育和研究中心之一。达喀尔也是一个重要的艺术和文化中心，拥有非洲艺术博物馆和塞内加尔历史博物馆。该城市因举办各种文化活动而享誉盛名，如著名的达喀尔双年展和达喀尔电影节。早年的达喀尔拉力赛也使其闻名于世。通过开罗－达喀尔高速公路和非洲西部沿海横贯高速公路，使其与邻国交通便利。

Dakar, located on the Cape Verde Peninsula along the Atlantic coast, is the capital and largest city of Senegal. As the westernmost city on the African continent, it occupies a crucial position for transatlantic and European trade, making it an important seaport. The city is home to several universities and research institutions, including the University of Senegal, establishing it as one of the primary education and research centers in West Africa. Dakar is also an important center for arts and culture, housing the Museum of African Art and the Senegal Historical Museum. The city is renowned for hosting various cultural events and festivals, such as the famed Dakar Biennale and Dakar Film Festival. Its earlier Dakar Rally also brought it global fame. Dakar is well-connected to neighboring countries through the Cairo-Dakar Highway and the Trans-West African Coastal Highway.

马里 巴马科

Bamako, Mali

　　巴马科是马里的首都，坐落于尼日尔河南岸，是该国最大的城市和主要的经济、文化、政治中心。国家政府部门、驻马里的各国使馆和联合国机构均设在此处。巴马科是马里的主要港口，尼日尔河作为该城市重要的交通枢纽，使巴马科成为贸易和运输的中心。在教育领域，巴马科拥有多所大学和学院，其中马里大学是西非最早的高等教育机构之一。文化方面，巴马科是马里的艺术和文化中心，重要的文化机构包括马里国家博物馆。在建筑上，巴马科具有法式风格、伊斯兰风格和传统非洲风格的建筑。此外，尼日尔河的丰富水资源使巴马科周边地区成为了马里的主要农业区，主要作物是小米、玉米和棉花。

Bamako, the capital of Mali, is situated on the southern bank of the Niger River. It serves as the country's largest city and principal economic, cultural, and governmental center. National government departments, foreign embassies in Mali, and United Nations agencies are all headquartered here. As the main port of Mali, Bamako plays a vital role in trade and transportation, with the Niger River offering a crucial transportation link for the city. In the field of education, Bamako is home to several universities and colleges, including the University of Mali, one of the earliest institutions of higher education in West Africa. Culturally, Bamako is the artistic and cultural center of Mali, with major cultural institutions like the National Museum of Mali. The cityscape showcases architectural styles from the French , along with Islamic and traditional African architectural arts. Moreover, the abundant water resources of the Niger River make the areas surrounding Bamako a major agricultural region in Mali, with primary crops including millet, corn, and cotton.

马 里

巴马科

km
0　1　2　3　4

几内亚 科纳克里

Conakry, Guinea

科纳克里是几内亚的首都和最大城市，几内亚的经济、金融和文化中心。科纳克里最初坐落于汤博岛上，后来扩展到邻近的卡卢姆半岛，中间有堤道和铁路连接。科纳克里的经济围绕其港口发展，该港口具备现代化的货物装卸和储存设施，主要运输产品为氧化铝和香蕉。本地的制造业包括食品加工、水泥制造、金属制品和燃料产品。科纳克里是索尼阿里大学的所在地，该校是几内亚最重要的高等教育机构。此外，几内亚国家博物馆和萨达纳广场也坐落于此，它们是科纳克里最受游客欢迎的旅游景点。

Conakry, the capital and largest city of Guinea, serves as the economic, financial, and cultural hub of the country. Initially located on the Tombo Island, Conakry later expanded onto the adjacent Kaloum Peninsula, connected by a causeway and railway. The economy of Conakry primarily revolves around its port, which possesses modern cargo handling and storage facilities, mainly transporting alumina and bananas. The local manufacturing sector includes food processing, cement, metal goods, and fuel products. Conakry is home to Soni Ali University, the country's most prestigious institution of higher education. Additionally, the National Museum of Guinea and Sardana Square, both located in the city, are among Conakry's most popular tourist attractions.

几内亚

科纳克里

km

0   1   2   3   4

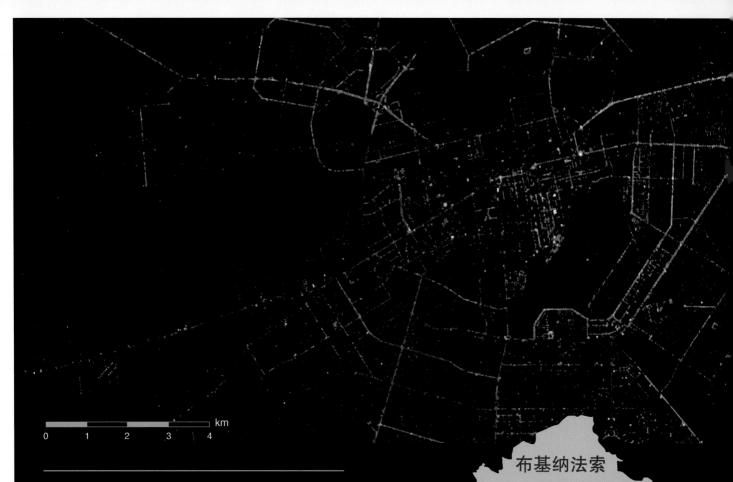

布基纳法索 瓦加杜古
Ouagadougou, Burkina Faso

布基纳法索

瓦加杜古

　　瓦加杜古是布基纳法索的首都，位于该国中部，是该国最大的城市以及经济、文化和政治中心。国家的所有重要政府机构和部门、各国驻布基纳法索的使馆和国际组织设在此地。瓦加杜古市区被划分为多个区，有新旧两个市中心，其中新市中心是行政和商业区，旧市中心则包含大部分的住宅和市场。瓦加杜古经济以服务业为主，包括银行、保险、零售和酒店，同时是布基纳法索最大的制造中心，主要制造业有食品加工和纺织。在教育方面，瓦加杜古拥有布基纳法索大学，是该国最重要的高等教育机构。此外，瓦加杜古还拥有布基纳法索国家图书馆和布基纳法索国家档案馆，是该国的重要学术和研究中心。在文化活动方面，瓦加杜古每年都会举办非洲电影节，这是非洲最重要的电影节之一。

Ouagadougou, the capital of Burkina Faso, is located in the center of the country and serves as its largest city and the main economic, cultural, and governmental hub. All significant government departments and institutions, foreign embassies in Burkina Faso, and international organizations are located here. The city is divided into several districts, with new and old city centers; the new center hosts administrative and business activities, while the old center comprises the majority of residential areas and markets. Ouagadougou's economy is primarily service-oriented, including banking, insurance, retail, and hotels, and it is also the largest manufacturing center in Burkina Faso, with the main industries being food processing and textiles. In education, Ouagadougou is home to the University of Burkina Faso, the most important institution of higher education in the country. Additionally, Ouagadougou houses the National Library and National Archives of Burkina Faso, making it an important academic and research center. Culturally, Ouagadougou annually hosts the African Film Festival, one of the most important film festivals in Africa.

科特迪瓦　亚穆苏克罗
Yamoussoukro, Côte d'Ivoire

亚穆苏克罗是科特迪瓦的首都。亚穆苏克罗位于阿比让西北 240 km 处，面积为 2075 km²。亚穆苏克罗是科特迪瓦的交通枢纽，拥有国际机场并有通向内陆的铁路网。城市中拥有多所重要的教育机构，其中最引人注目的是科特迪瓦大学，它是全国规模最大且历史最悠久的大学。亚穆苏克罗的建筑风格独特，吸引了大量游客，其中和平圣母大教堂尤为出名。亚穆苏克罗地处赤道气候带，气候炎热湿润，雨量充沛。年降水量在 1300 ~ 2200 mm 之间，平均温度在 24 ~ 27℃之间。

Yamoussoukro is the capital of Côte d'Ivoire. Located approximately 240 km northwest of Abidjan, it spans an area of 2075 km². Yamoussoukro serves as a transportation hub in Côte d'Ivoire, boasting an international airport and a railway network that connects to the interior. The city is home to many important educational institutions, notably the University of Côte d'Ivoire, which is the largest and oldest university in the country. The unique architecture of Yamoussoukro attracts a large number of tourists, especially the Basilica of Our Lady of Peace. Yamoussoukro is situated in the equatorial climate zone, characterized by hot and humid conditions with abundant rainfall. The annual precipitation ranges from 1300 to 2200 mm, with average temperatures ranging from 24℃ to 27℃.

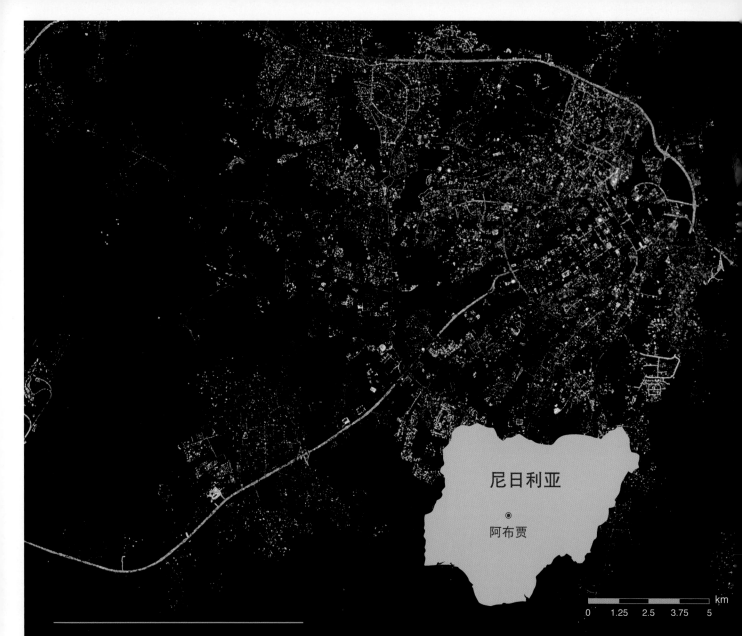

尼日利亚

⊙
阿布贾

km
0   1.25   2.5   3.75   5

尼日利亚 阿布贾

Abuja, Nigeria

　　阿布贾是尼日利亚首都,位于中部尼日尔河支流古拉河畔,联邦首都区面积 7315 km²。该城市原为国家锡矿开采地、中部公路网中心和农畜产品集散地,为加强联邦政府与各地区、各民族之间的联系,并解决原首都拉各斯规模过大所带来的问题,1976 年决定迁都于此,并划定联邦首都区范围。阿布贾为热带草原气候,凉爽宜人,水资源丰富。市区规划为新月形,位于瓜瓜平原上。

　　Abuja is the capital of Nigeria, situated in the central region along the banks of the Gurara River, a tributary of the Niger River. The Federal Capital Territory covers an area of 7315 km². Originally, this city was a site for tin mining, a central hub for road networks, and a center for the distribution of agricultural and livestock products. In a bid to strengthen the connections between the federal government and various regions and ethnic groups, as well as to address the challenges posed by the excessive size of the former capital, Lagos, a decision was made in 1976 to relocate the capital to Abuja and define the boundaries of the Federal Capital Territory. Abuja features a tropical grassland climate, offering a pleasant and cool environment with abundant water resources. The city is planned in a crescent shape and is located on the Gwagwa Plain.

赤道几内亚 马拉博
Malabo, Equatorial Guinea

马拉博，1973 年之前称圣伊萨贝尔，是赤道几内亚的首都和最大城市，坐落在国内最大岛屿比奥科岛的北部。作为海岸城市，马拉博享有港口城市的优势，已经发展为全国最大的进出口贸易港口，并设有航线通往木尼河地区以及其他西非国家。马拉博有一些令人瞩目的建筑，其中包括巴洛克风格的赤道几内亚总统府，以及圣伊莎贝尔大教堂和银行大厦。马拉博的经济以石油和天然气工业为主，这也使得该城市成为赤道几内亚最重要的经济中心。除石油和天然气产业外，渔业和农业也是马拉博的主要经济支柱。由于历史和地理因素，马拉博有多种官方语言，包括西班牙语、法语、葡萄牙语和一些原住民语言。

Malabo, known as Santa Isabel until 1973, is the capital and the largest city of Equatorial Guinea, located in the north of the country's largest island, Bioko Island. As a coastal city, Malabo enjoys the advantages of a port city and has developed into the largest import and export trade port in the country, with air routes to the mainland's Muni River Region and other West African countries. Malabo has some eye-catching buildings, including the Baroque-style Presidential Palace of Equatorial Guinea, as well as the Cathedral of Santa Isabel and Bank Building. The economy of Malabo is primarily driven by the oil and natural gas industry, making the city the most important economic center in Equatorial Guinea. Besides the oil and natural gas industry, fishing and agriculture are also major economic pillars in Malabo. Due to historical and geographical factors, multiple official languages are used in daily life in Malabo, including Spanish, French, Portuguese, and some indigenous languages.

开罗

埃　　及

埃及 开罗
Cairo，Egypt

开罗，作为埃及的首都，人口超过 1000 万，是该国最大的城市，也是非洲、阿拉伯世界和中东最大的城市，还是整个中东地区的政治、经济、文化、商业和交通中心，影响力渗透到了整个阿拉伯世界。这座城市与古埃及文明紧密相连，无论是吉萨的金字塔群，还是太阳城赫里奥波里斯，都是开罗历史的象征，同时也是吸引全球游客的热门景点。开罗因其丰富的伊斯兰建筑被誉为"千塔之城"，既彰显了城市的历史沉淀，也为其现代化的天际线增添了独特韵味。作为阿拉伯世界历史最悠久、规模最大的电影和音乐产业中心，开罗对阿拉伯世界的流行文化产生了深远影响。开罗的爱资哈尔大学是世界上历史最悠久的高等学府之一，对于培养阿拉伯世界的精英人才起着重要的作用。

Cairo, serving as the capital of Egypt and the core of the Cairo Governorate, boasts a population of over 10 million, making it not only the largest city in the country, but also a part of the largest urban conglomeration in Africa, the Arab world, and the Middle East. Cairo represents the political, economic, cultural, commercial, transportation hub center of throughout the entire Middle East region, with its influence permeating the entire Arab world. The city's history is inextricably intertwined with that of ancient Egypt, with landmarks such as the Giza Pyramids, and the ancient cities of Heliopolis, all symbolizing Cairo's historical legacy and serving as popular destinations for tourists worldwide. Known as the "City of a Thousand Minarets" for its abundance of Islamic architecture, these buildings both manifest the city's historical depth and add a unique flavor to its modern skyline. As the oldest and largest center of the film and music industry in the Arab world, Cairo has profoundly influenced Arab popular culture. Al-Azhar University in Cairo, one of the oldest university in the world, plays a crucial role in cultivating the elite talents of the Arab world.

苏 丹

喀土穆

苏丹 喀土穆
Khartoum, Sudan

　　喀土穆是苏丹的首都，始建于 1822 年，位于白尼罗河和青尼罗河的交会处。尼罗河从这里继续向北流向埃及和地中海。喀土穆被尼罗河分成三部分，城市人口超过 500 万。喀土穆是北非的经济和贸易中心，与苏丹港和欧拜伊德铁路相通。喀土穆市区有多个国家和文化机构，包括苏丹国家博物馆、喀土穆大学和苏丹科技大学。

　　Khartoum is the capital of Sudan and was founded in 1822. It is located at the confluence of the White Nile and the Blue Nile. From there, the Nile continues north towards Egypt and the Mediterranean Sea. Divided by these two parts of the Nile, Khartoum is a tripartite metropolis with an estimated population of over five million people, consisting of Khartoum proper. Khartoum is an economic and trade center in Northern Africa, with rail lines from Port Sudan and El-Obeid. Several national and cultural institutions are in Khartoum and its metropolitan area, including the National Museum of Sudan, the Khalifa House Museum, the University of Khartoum, and the Sudan University of Science and Technology.

吉布提 吉布提
Djibouti City, Republic of Djibouti

　　吉布提市是东非国家吉布提的首都，同时也是该国人口最多、面积最大的城市，位于该国的东部。得益于其紧邻红海和印度洋的优越地理位置，吉布提市已经发展为该地区的重要海港和国际航运中心，为中东、非洲和欧亚之间的航运提供了便利，并且是邻国埃塞俄比亚的主要海运出口港口。吉布提市是非洲之角的第二大经济体，被誉为"塔朱拉湾的明珠"。作为吉布提的工业中心，吉布提市的主要行业包括食品加工和建筑材料生产。吉布提市也是吉布提教育中心，拥有该国最大的高等教育机构——吉布提大学。

Djibouti City is the capital and the largest city of the East African country, Djibouti, located in the eastern part of the country. Thanks to its strategic location near the Red Sea and the Indian Ocean, Djibouti City has developed into a significant seaport and an international shipping center in the region, providing convenience for shipping between the Middle East, Africa, and Eurasia. Furthermore, it serves as the primary seaport for exports from its neighbor, Ethiopia. Djibouti City is the second-largest economy in the Horn of Africa and is known as the "Pearl of the Tadjoura Gulf". As the main industrial center of Djibouti, the city's key industries include food processing and building materials production. Djibouti City is also the primary education center of Djibouti, home to the country's largest higher education institution — the University of Djibouti.

吉布提

吉布提

```
                          km
0      1      2      3      4
```

乌干达 坎帕拉
Kampala, Uganda

坎帕拉，乌干达的首都和最大城市，坐落在乌干达南部，毗邻维多利亚湖。这座城市被划分为五个政治区，是乌干达的经济、政治和文化中心，也是非洲人口增长最快的城市之一，其年人口增长率达到4.03%。坎帕拉的主要产业包括制造业、服务业和农业，是非洲最大的移动电话市场之一。在教育方面，坎帕拉拥有马凯雷大学，这是东非最古老的大学，成立于1922年。在文化方面，这里有东非最古老的博物馆——乌干达国家博物馆，以及非洲艺术中心，该中心致力于推广非洲艺术和文化。作为乌干达的交通枢纽，这座城市拥有国际机场和大型公路网络，使其成为乌干达的主要入口。坎帕拉被公认为东非最适合居住的城市。

Kampala, the capital and largest city of Uganda, is located in the southern part of the country, adjacent to Lake Victoria. The city is divided into five political divisions, serving as the economic, political, and cultural center of Uganda. It's also one of Africa's fastest-growing cities with an annual population growth rate reaching 4.03%. Kampala's main industries include manufacturing, services, and agriculture and it's one of the largest mobile phone markets in Africa. In terms of education, Kampala is home to Makerere University, the oldest university in East Africa, established in 1922. In culture, it hosts the oldest museum in East Africa — the Uganda National Museum, and the African Art Center, an institution dedicated to promoting African art and culture. As a transport hub in Uganda, the city has an international airport and a large highway network, making it a major entry point to Uganda. Kampala is recognized as the most livable city in East Africa.

乌干达

◉
坎帕拉

km
0   1.25   2.5   3.75   5

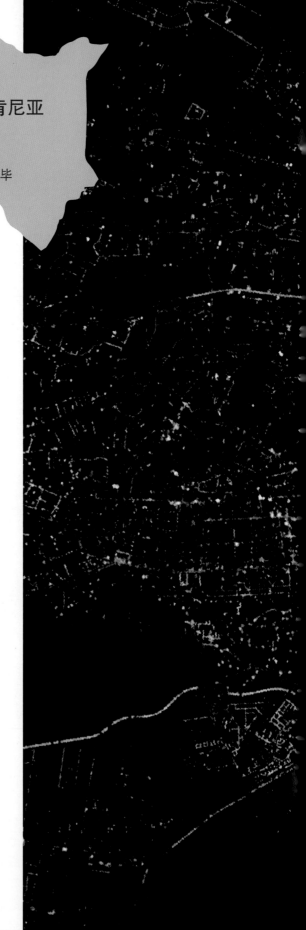

肯尼亚 内罗毕
Nairobi, Kenya

　　内罗毕是肯尼亚的首都，人口约 300 万，得名于穿城而过的内罗毕河，是非洲的大城市之一。尽管位于赤道附近，然而其海拔 1660 m 的高原地理位置使得它拥有与赤道附近其他炎热地区迥然不同的凉爽气候。在历史上，它曾是咖啡、茶叶和剑麻工业的重要中心。内罗毕不仅聚集了数千家肯尼亚企业，还吸引超过一千家国际公司和组织，其中包括联合国环境规划署和联合国内罗毕办事处。作为一个成熟的商业和文化中心，内罗毕拥有非洲最大的证券交易所之一，并于 2010 年加入了联合国教科文组织的全球学习型城市网络。此外，该城市还以内罗毕国家公园而闻名。

　　Nairobi, the capital of Kenya, with a population of about three million, is named after the Nairobi River that runs through the city. It is also one of the major cities in Africa. Despite being located near the equator, its high-altitude location of 1660 m gives it a cool climate that is different from other hot areas near the equator. In history, it became an important center for coffee, tea and sisal industries. Nairobi is home to the Kenyan Parliament House and attracts thousands of Kenyan businesses as well as over a thousand international companies and organizations, including the United Nations Environment Programme and the United Nations Office at Nairobi. As a mature commercial and cultural center, Nairobi has one of the largest stock exchanges in Africa and joined the UNESCO Global Network of Learning Cities in 2010. In addition, the city is also famous for its Nairobi National Park.

肯尼亚 蒙巴萨
Mombasa, Kenya

肯尼亚

蒙巴萨

    蒙巴萨坐落于肯尼亚东南部，东临印度洋。作为肯尼亚历史最悠久的城市之一，蒙巴萨以其繁荣的历史海港而著名，是肯尼亚第二大城市，仅次于内罗毕。蒙巴萨拥有肯尼亚最大的海港，也是东非地区最大的国际贸易港口之一，对肯尼亚和整个东非地区的经济发展至关重要。每年都有大量的货物在这里装卸，为肯尼亚的经济发展做出了巨大贡献。这座城市的历史可以追溯到公元 900 年左右，是印度洋重要贸易中心之一。由于其战略地位的重要性，历史上曾多次成为争夺对象，被多个国家争相占领。如今，以旅游业为主导的蒙巴萨，拥有国家大厦、大型港口和国际机场，展现出一片充满活力和希望的景象。

Mombasa is located in the southeast of Kenya, facing the Indian Ocean. As one of the oldest cities in Kenya, Mombasa is famous for its prosperous historical port and is the second largest city in Kenya after Nairobi. Mombasa has the largest port in Kenya and is one of the largest international trade ports in East Africa, which is vital for the economic development of Kenya and the entire East African region. Every year, a large number of goods are loaded and unloaded here, making a huge contribution to Kenya's economic development. The history of this city can be traced back to around 900 AD and was one of the important trade centers on the Indian Ocean. Due to its strategic importance, it was historically contested by many countries and occupied by several powers. Today, Mombasa is dominated by tourism and has national buildings, large ports and international airports, showing a vibrant and hopeful scene.

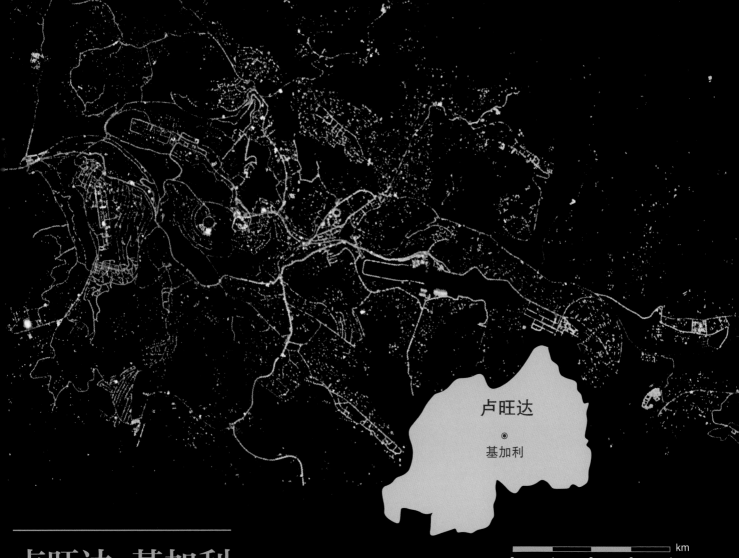

卢旺达

基加利

km
0 1 2 3 4

卢旺达 基加利

Kigali, Rwanda

基加利位于卢旺达的地理中心，海拔高度在 1300 ～ 1600 m 之间，被壮丽的山脉环绕，享有宜人的凉爽气候。作为卢旺达的首都和最大城市，基加利自 1962 年卢旺达独立以来，一直是国家的经济、文化和交通枢纽。基加利拥有现代化的国际机场，提供前往非洲其他国家以及欧洲和亚洲的航班。基加利的国内生产总值主要贡献来自服务业，包括金融、保险和房地产等。受益于卢旺达农业的发展，基加利成为农产品交易的重要中心。基加利被誉为非洲最干净、最安全的城市之一，这归功于卢旺达政府的有效治理。此外，该城市还致力于发展旅游业，吸引国际游客，其中，基加利纪念馆是基加利的重要景点，每年都会吸引大量游客前来参观。

Kigali is located in the geographical center of Rwanda, with an altitude ranging from 1300 to 1600 m. It is surrounded by magnificent mountains and enjoys a pleasant cool climate. As the capital and largest city of Rwanda, Kigali has been the economic, cultural and transportation hub of the country since Rwanda's independence in 1962. It has a modern international airport that offers flights to other African countries as well as Europe and Asia. Kigali's gross domestic product mainly comes from the service sector, including finance, insurance and real estate. Kigali has become an important center for agricultural trade thanks to Rwanda's agricultural development. Kigali is known as one of the cleanest and safest cities in Africa, thanks to the effective governance of the Rwandan government. In addition, the city is also committed to attracting international tourists, including leisure tourism, conferences and exhibitions. Among them, the Genocide Memorial is an important attraction in Kigali that attracts a large number of visitors every year.

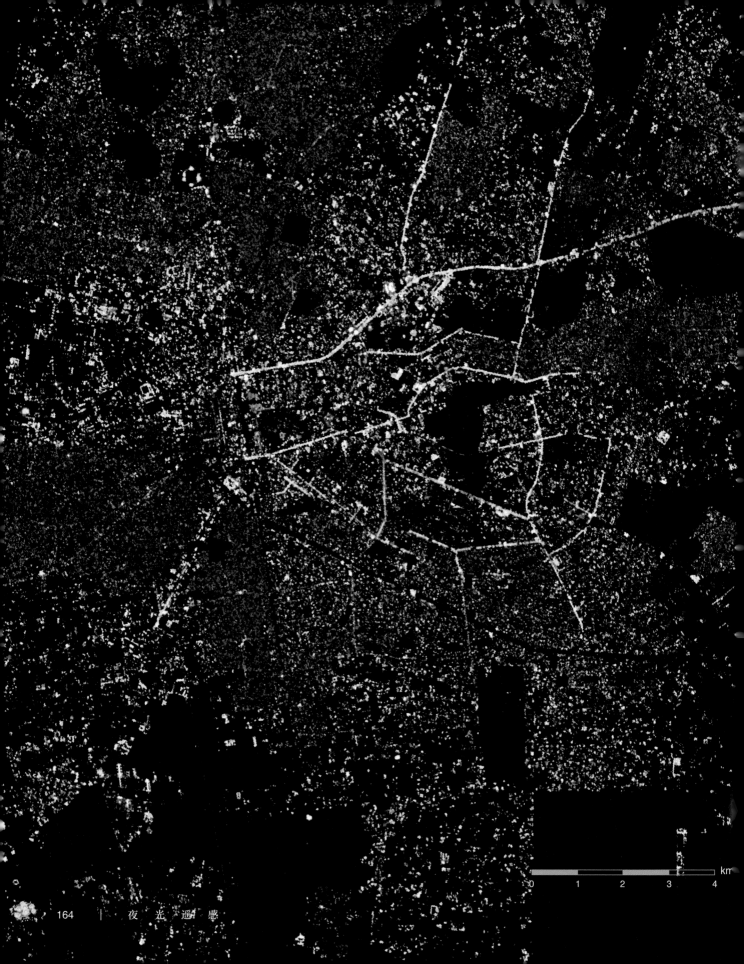

赞比亚

卢萨卡 ◎

赞比亚 卢萨卡
Lusaka, Zambia

　　卢萨卡是赞比亚的首都和第一大城市，非洲南部发展最快的城市之一，坐落于赞比亚中部高原的南部，海拔约 1279 m。卢萨卡不仅是赞比亚的商业和政治中心，还是该国的交通枢纽，连接着东南西北四个方向的高速公路。在 1964 年赞比亚独立后，卢萨卡成为了新国家的首都。随后，该城市开展了大规模的建设规划，兴建了政府大楼、赞比亚大学以及机场等基础设施，为卢萨卡的发展打下了坚实的基础。

　　Lusaka, as one of the fastest growing cities in southern Africa, is the capital and largest city of Zambia. It is located in the southern part of Zambia's central plateau at an altitude of about 1279 m. Lusaka is not only Zambia's commercial and political center but also connects four major highways leading to north-south-east-west directions in Zambia. After Zambia's independence in 1964, Lusaka became capital again for new nation. Subsequently large-scale construction plans were carried out for building government buildings University of Zambia airport etc which laid solid foundation for Lusaka's development.

欧洲

EUROPE

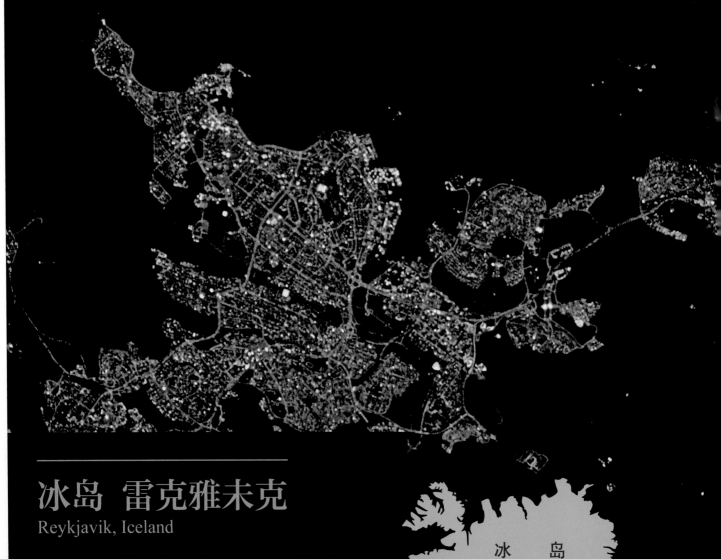

冰岛 雷克雅未克
Reykjavik, Iceland

冰 岛

◎ 雷克雅未克

km
0　　　1.25　　　2.5　　　3.75　　　5

雷克雅未克是冰岛的首都和最大的城市，同时也是全国文化、经济和政治中心。雷克雅未克位于冰岛西部法赫萨湾的东南角，处于北纬 64°08′的高纬上，是全球最北的国家首都，占地 273 km²。雷克雅未克的对外交通设施以公路为主，工业能源以温泉的地热能为主，其温泉水即使经过管道输送后，水温仍高达 90℃以上，故此雷克雅未克极少使用石油和煤等能源，城市环境受污染程度极低，有"无烟城"的美誉。城市人口约为24 万人。

Reykjavik is the capital and largest city of Iceland, as well as the center of culture, economy and politics of the country. It is located in the southeast corner of Faxa Bay in western Iceland, with a latitude of 64°08′N, making it the northernmost national capital in the world. It covers an area of 273 km². Reykjavik's external transportation facilities are mainly centered around highways, and its industrial energy primarily relies on the geothermal energy from hot springs. Even after being transported through pipelines, the temperature of the geothermal water remains above 90℃. As a result, Reykjavik rarely utilizes energy sources such as oil and coal, leading to minimal pollution levels in the urban environment. The city is renowned as a smoke-free city. The population of the city is about 240000.

挪威 奥斯陆
Oslo, Norway

　　奥斯陆是挪威的首都，也是该国最大和人口最多的城市，位于挪威东南部，其面积约为 454 km²，是挪威的政治、经济、文化、交通中心和主要港口，也是挪威王室和政府的所在地。奥斯陆是全欧洲最富有、最安全和拥有最高生活水准的城市之一，也是世界上最幸福的城市之一，还是 1952 年奥斯陆冬季奥运会的主办城市。奥斯陆也是诺贝尔和平奖的颁奖地，每年的颁奖仪式就在奥斯陆市政厅举行。根据考古学研究，奥斯陆于 1000 年前后建城。奥斯陆老城区是北欧除了维斯比以外最大的中世纪城市，仍然保存完好。由于奥斯陆三面被群山、丛林和原野所环抱，城市街道两旁的建筑大多只有六七层，带有浓厚的中世纪色彩和独具一格的北欧风光。苍山与绿原相辉映，风景十分迷人。

　　Oslo is the capital of Norway and also its largest and most populous city. It is located in the southeastern part of Norway, covering an area of approximately 454 km². It serves as the political, economic, cultural, and transportation hub of Norway, as well as its main port. Oslo is also the residence of the Norwegian royal family and the government. It stands as one of the wealthiest, safest, and highest living standard cities in all of Europe, and is considered one of the happiest cities in the world. It was the host city for the 1952 Winter Olympics and the venue for the Nobel Peace Prize ceremony, held annually at Oslo City Hall. According to archaeological research, Oslo was founded around the year 1000. The Old Town of Oslo is one of the largest medieval cities in Northern Europe, and it remains well-preserved. Bordered by mountains, forests, and plains on three sides, Oslo's buildings along its streets typically rise only six or seven stories, exuding a strong medieval charm and a unique Northern European atmosphere. The juxtaposition of verdant fields and serene mountains creates an enchanting landscape.

挪

威

奥斯陆

km

0　　1.25　　2.5　　3.75　　5

丹 麦

哥本哈根。

丹麦 哥本哈根
Copenhagen, Denmark

　　哥本哈根是丹麦的首都，也是丹麦最大的城市，拥有该国最大的港口，城市的面积达 180 km²，位于西兰岛东部，是丹麦的政治、经济和文化中心。哥本哈根曾被联合国人居署选为"全球最宜居的城市"，并给予"最佳设计城市"的评价。哥本哈根既是传统的贸易和船运中心，又是新兴制造业城市，全国 1/3工厂建在大哥本哈根区。哥本哈根市容美观整洁，市内新兴的大工业企业和中世纪古老的建筑物交相辉映，使它既是现代化的都市，又具有古色古香的特色，是世界上著名的历史文化名城。丹麦标志美人鱼雕像在海边静静沉思，充满童话气质的古堡与皇宫比邻坐落在这个城市中，古老与神奇、艺术与现代。

　　Copenhagen is the capital of Denmark and also the largest city in the country, boasting its largest port. The city covers an area of 180 km² and is situated on the eastern part of Zealand Island. It serves as Denmark's political, economic, and cultural hub. Copenhagen has been selected by the United Nations as one of the "most livable cities globally" and has received praise for being the "best-designed city". Copenhagen is both a traditional trading and maritime center and an emerging manufacturing city. One-third of the nation's factories are situated in the Greater Copenhagen area. Copenhagen's urban landscape is characterized by its aesthetic cleanliness, where modern emerging industries coexist with medieval architecture, creating a blend of modernity and historical charm. This city is renowned worldwide as a historical and cultural landmark. The iconic Little Mermaid statue, symbolizing Denmark, contemplates by the sea, while fairy tale-like castles and palaces neighbor one another, infusing the city with an atmosphere of ancient wonder and artistic modernity.

德国 柏林
Berlin, Germany

柏林是德国首都，也是德国最大和人口最多的城市，城区面积达 3743 km²，其中该市约三分之一的面积由森林、公园、花园、河流、运河和湖泊组成。柏林不仅是文化、政治、媒体和科学的国际化都市，还以其高科技公司和服务业为基础的经济而闻名。此外，柏林还是欧洲大陆的航空和铁路交通枢纽，拥有高度复杂的公共交通网络。柏林的工业产能包括食品、电子产品和机械生产。这座城市的经济多元且充满活力，吸引了大量国内外企业的投资和发展。

Berlin is the capital of Germany and also the largest city in Germany in terms of area and population. Its urban area covers an area of 3743 km², of which about one-third of the city is composed of forests, parks, gardens, rivers, canals and lakes. Berlin is not only a cosmopolitan city of culture, politics, media and science, but also known for its economy based on high-tech companies and services. In addition, Berlin is also a major air and rail transport hub on the European continent, with a highly complex public transport network. Berlin's industrial capacity includes food, electronics and machinery production. The city has a diverse and vibrant economy that attracts a large number of domestic and foreign enterprises to invest and develop.

捷克 布拉格
Prague, Czech Republic

　　布拉格是捷克共和国的首都和最大的城市，同时也是中欧的政治、文化和经济中心，占地面积为496 km²。布拉格的主要产业包括机动车制造、软件开发、电子产品生产等。作为捷克的中心城市，布拉格在历史、文化和艺术方面具有重要地位。其古老的建筑、古迹和风景使其成为了独特的旅游胜地，吸引着来自世界各地的游客。

　　Prague is the capital and largest city of the Czech Republic, as well as the political, cultural and economic center of Central Europe. The city covers an area of 496 km². Prague's main industries include motor vehicle manufacturing, software development, electronics production and more. As the central city of the Czech Republic, Prague has an important position in history, culture and art. Its ancient buildings, monuments and scenery make it a unique tourist destination that attracts visitors from all over the world.

km

| 0 | 1.25 | 2.5 | 3.75 | 5 |

◎ 布拉格

捷 克

km
0   1.25   2.5   3.75   5

阿姆斯特丹

荷 兰

荷兰 阿姆斯特丹

Amsterdam, Netherlands

阿姆斯特丹是荷兰的首都和人口最多的城市，位于荷兰的北荷兰省。由于城市内有大量的运河，它被俗称为"北方的威尼斯"。这座港口城市整体地势较低，总体海拔约 2 m，并且部分城区是通过填海而建造的，城区的总面积达 219 km²。阿姆斯特丹是金融和贸易的主要中心，最早的证券交易市场和资本主义市场，诞生了全球第一家证券交易所，同时这座城市还因其丰富的夜生活和节日活动而闻名。著名的荷兰风车、木鞋、郁金香、奶酪等吸引了大量的游客。为满足各国游客的需求，荷兰人在阿姆斯特丹市附近建造了集民风民俗于一体的民俗村，受到游客欢迎。

Amsterdam is the capital and most populous city of the Netherlands, located in the province of North Holland. Due to the large number of canals in the city, it is colloquially known as "the Venice of the North". The overall terrain of this port city is low-lying, with an average elevation of 2 m. Some parts of the city were built by reclamation from the sea. The urban area covers an area of 219 km². Amsterdam is a major center for finance and trade, home to the world's first stock exchange and capitalism market. It is renowned for its vibrant nightlife and festive events. The globally famous Dutch windmills, wooden shoes, tulips, cheese, and more constitute several major cultural attractions that draw tourists. To cater to the needs of visitors from various countries, the Dutch have established folk villages near Amsterdam, integrating culture and customs, which have been well-received by tourists.

爱尔兰 都柏林
Dublin, Ireland

　　都柏林是爱尔兰的首都和最大城市，坐落在利菲河口的一个海湾上，位于爱尔兰东部伦斯特省，城区面积为 318 km²。都柏林与都柏林山脉的南部接壤，同时也是威克洛山脉的一部分。作为爱尔兰的重要城市，都柏林是教育、艺术、文化、行政和工业中心。都柏林不仅以其历史悠久的文化遗产和艺术场所而闻名，还是现代产业和创新的重要中心之一。

　　Dublin is the capital and largest city of Ireland, situated on a bay at the mouth of the River Liffey in the province of Leinster in eastern Ireland, with covering an area of 318 km². It borders the south of the Dublin Mountains, which are also part of the Wicklow Mountains. As an important city in Ireland, Dublin is the center of education, art, culture, administration and industry. Dublin is not only famous for its long-standing cultural heritage and artistic venues, but also one of the important centers for modern industry and innovation.

km
0　　1.25　　2.5　　3.75　　5

都柏林 ◉

爱 尔 兰

km
1.25   2.5   3.75   5

◉ 布鲁塞尔

比利时

比利时 布鲁塞尔
Brussels, Belgium

　　布鲁塞尔是比利时的首都，位于塞纳河畔，北部是低平的弗兰德平原，南部是略有起伏的布拉邦特台地，平均海拔 58 m，城区面积 161 km²，是布鲁塞尔首都地区的历史中心。除此之外，布鲁塞尔还是欧盟的行政中心，拥有许多主要的欧盟机构，200 多个国际行政中心及超过 1000 个官方团体的日常会议举办城市，有"欧洲首都"之称。布鲁塞尔拥有全欧洲最精美的建筑和博物馆，摩天大楼跟中世纪古建筑相得益彰。整座城市以皇宫为中心，沿"小环"而建，游览以步行为佳。布鲁塞尔是一个双语城市，通用法语和荷兰语，法语使用者占较多数。另外土耳其语、阿拉伯语等语言也被布鲁塞尔的穆斯林广泛使用。

　　Brussels is the capital of Belgium, located on the banks of the Senne River. To the north lies the flat Flemish Plain, while the slightly undulating Brabant Plateau is to the south, with an average elevation of 58 meters. The city covers an area of 161 square kilometers and is the historical center of the Brussels-Capital Region. Moreover, Brussels serves as the administrative center of the European Union, housing many major EU institutions. It is a city where over 200 international administrative centers and more than 1000 official groups hold daily meetings, earning it the title of the "Capital of Europe". Brussels boasts some of the most exquisite architecture and museums in all of Europe, with skyscrapers harmoniously blending with medieval structures. The city is centered around the Royal Palace and is built along the "Small Ring", making it perfect for exploration on foot. Brussels is a bilingual city, with both French and Dutch being widely spoken. French is the more predominant language. Additionally, languages like Turkish and Arabic are also extensively used by the Muslim community in Brussels.

km
0   1.25   2.5   3.75   5

英国 伦敦
London, England

伦敦是英国的首都和最大的城市，面积约为 1572 km²。在艺术、娱乐、商业、教育、医疗、科技和旅游等方面，伦敦都具有强大的影响力。伦敦以其丰富的文化氛围、多样化的活动和各种产业而著名。伦敦的机场系统以及城市公共交通网络以其繁忙和复杂程度而闻名世界。作为一个多元化和国际化的城市，伦敦拥有众多的历史古迹、博物馆、艺术画廊、剧院和音乐场所，吸引了大量的游客和居民。伦敦也是全球金融中心之一，拥有众多国际金融机构和公司的总部。

London is the capital and largest city of the United Kingdom, with covering an area of about 1572 km². It has a strong influence in various fields such as art, entertainment, business , education, health, media, technology, and tourism. London is famous for its rich cultural atmosphere, diverse activities and various industries. London's airport system and urban public transport network are known worldwide for their busyness and complexity. As a diverse and international city, London has many historical monuments, museums, art galleries, theatres and music venues that attract a large number of tourists and residents. London is also one of the global financial centers, with headquarters of many international financial institutions and companies.

卢森堡

◉ 卢森堡

卢森堡 卢森堡
Luxembourg City

卢森堡的城市面积为 52 km², 是卢森堡大公国的首都, 全市人口来自 160 个民族, 外国人占该市人口的 70%, 而卢森堡本国居民占 30%, 外国出生居民数量每年稳步增长。卢森堡已发展成为银行和行政中心, 吸引了众多国际金融机构和公司。此外, 卢森堡也是欧盟行政中心之一, 与布鲁塞尔、法兰克福和斯特拉斯堡并列。

Luxembourg covers an area of 52 km² and is the capital of the Grand Duchy of Luxembourg. The city's population comes from 160 nationalities, with foreigners accounting for 70% of the city's population and Luxembourg nationals accounting for 30%. The number of foreign-born residents in the city grows steadily every year. Luxembourg has developed into a banking and administrative center, attracting the establishment of many international financial institutions and companies. Additionally, Luxembourg is also one of the administrative centers of the European Union, alongside Brussels, Frankfurt, and Strasbourg.

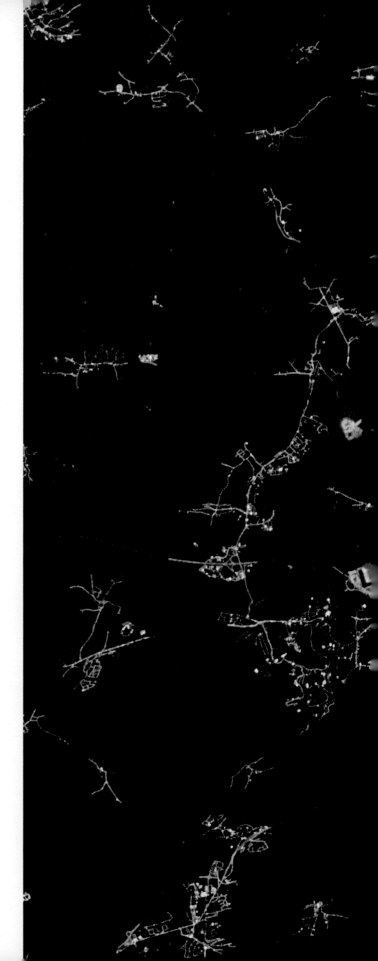

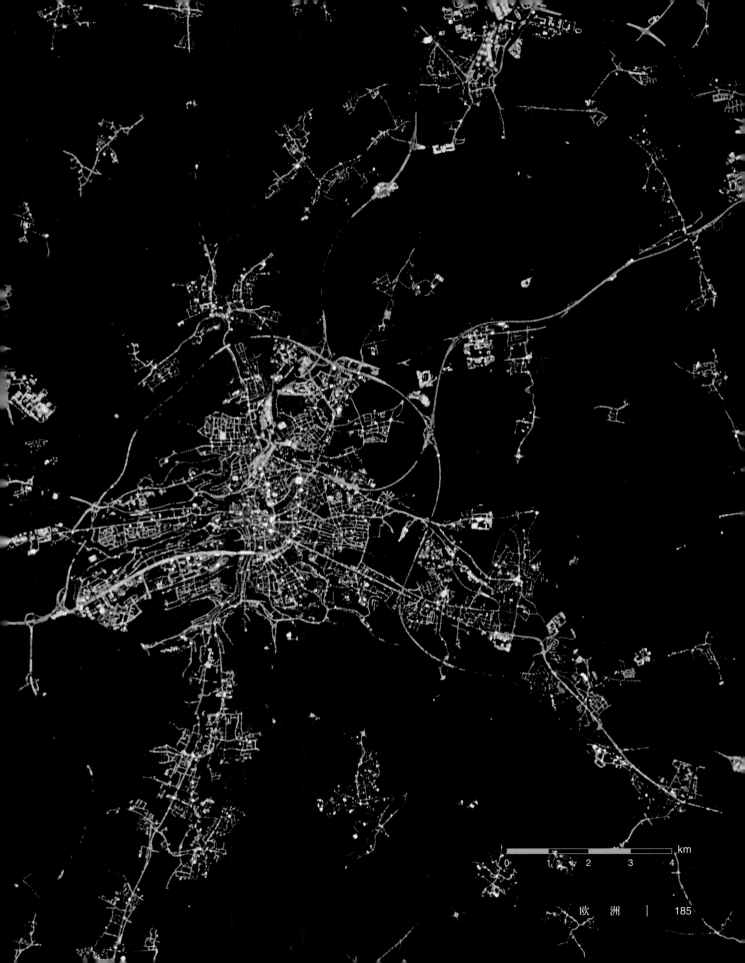

法国 巴黎
Paris, France

　　巴黎是法国的首都和人口最多的城市，城区面积达到 2800 km²。历史上一直是欧洲金融、外交、商业和科学的主要中心之一。作为法国的政治和文化中心，巴黎在全球范围内享有很高的声誉。巴黎以其独特的文化遗产、艺术、建筑和历史名胜而闻名，吸引了大量的游客和学者。巴黎还是世界上重要的时尚和美食之都，拥有各种高端时装品牌和美食餐厅。城市内有许多知名的博物馆、艺术画廊和历史古迹，如卢浮宫、埃菲尔铁塔、巴黎圣母院等。

Paris is the capital and most populous city of France which covers an area of 2800 km², and has historically been one of the main centers of finance, diplomacy, commerce and science in Europe. As the political and cultural center of France, Paris enjoys a high reputation worldwide. It is famous for its unique cultural heritage, art, architecture and historical attractions, attracting a large number of tourists and scholars. Paris is also one of the world's important fashion and gastronomy capitals, with various high-end fashion brands and gourmet restaurants. The city has many famous museums, art galleries and historical monuments, such as the Louvre Museum, the Eiffel Tower, Notre Dame Cathedral and more.

瑞士 伯尔尼

Bern, Switzerland

　　伯尔尼位于瑞士西部，被称为"联邦城市"，是瑞士的首都。穿城而过的阿勒河将伯尔尼老城三面环绕，这条河也将伯尔尼分为两个部分，西岸为老城区，东岸为新城区。这座历史悠久的城市有着丰富的文化遗产和建筑，吸引着游客和居民，早在 1983 年，旧城区伯尔尼古城的古钟塔就被联合国教科文组织评为世界历史文化遗产。作为瑞士的政治中心，伯尔尼承载着重要的行政和政府职能，同时也是国际外交活动的重要场所。伯尔尼也以"表都"著称，钟表商店比比皆是，即使到了郊外山乡小镇，也随处可见装潢雅致的钟表铺，走在伯尔尼大街上，犹如漫游在钟的海洋，表的世界，到处都是醒目的钟表广告。

　　Bern is located in western Switzerland and is known as the "Federal City", as it is the de facto capital of Switzerland. The Aare River runs through the city and surrounds Bern's old town on three sides. The river also divides Bern into two parts: the west bank is the old town area and the east bank is the new town area. This historic city has a rich cultural heritage and architecture that attracts tourists and residents. As early as 1983, the Old Town of Bern, known as the "Zytglogge", was designated as a UNESCO World Heritage site. As Switzerland's political center, Bern carries significant administrative and governmental functions, while also serving as an important hub for international diplomatic activities. Bern is also renowned as the "City of Clocks", with watch shops abound. Even in the small towns and villages in the outskirts and mountains, you can find elegantly decorated watch stores. Walking through the streets of Bern feels like wandering in a sea of clocks, a world of timepieces, with eye-catching watch advertisements everywhere.

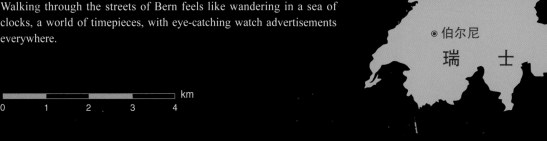

km

0　　1　　2　　3　　4

⊚ 伯尔尼

瑞　士

斯洛文尼亚 卢布尔雅那
Ljubljana, Slovenia

斯洛文尼亚
卢布尔雅那

卢布尔雅那是斯洛文尼亚的首都和最大的城市，同时也是全国的文化、教育、经济、政治和行政中心。作为斯洛文尼亚的首都，卢布尔雅那拥有丰富的历史和文化遗产，以及各种教育和研究机构。卢布尔雅那是一个充满活力的城市，拥有多样化的艺术、文化和娱乐活动。卢布尔雅那的市区以其宜人的氛围和风景如画的河岸景色而闻名。该城市自古就是交通要道，是斯洛文尼亚通往意大利、奥地利、巴尔干诸国的国际铁路枢纽。该城市是世界上最早在市区的街道上安装供暖设备的城市，在冬天的街道上看不到冰雪痕迹。

Ljubljana is the capital and largest city of Slovenia, as well as the national center of culture, education, economy, politics and administration. As the capital of Slovenia, Ljubljana has a rich history and cultural heritage, as well as various educational and research institutions. It is a vibrant city with diverse artistic, cultural and entertainment activities. Ljubljana's urban area is known for its pleasant atmosphere and picturesque riverside scenery. Ljubljana's city center is renowned for its pleasant atmosphere and picturesque riverside views. The city has long been a transportation hub, serving as an international railway junction connecting Slovenia to Italy, Austria, and the Balkans. Ljubljana was one of the first cities in the world to install heating systems on its streets, with no signs of snow and ice in winter.

km
0   1   2   3   4

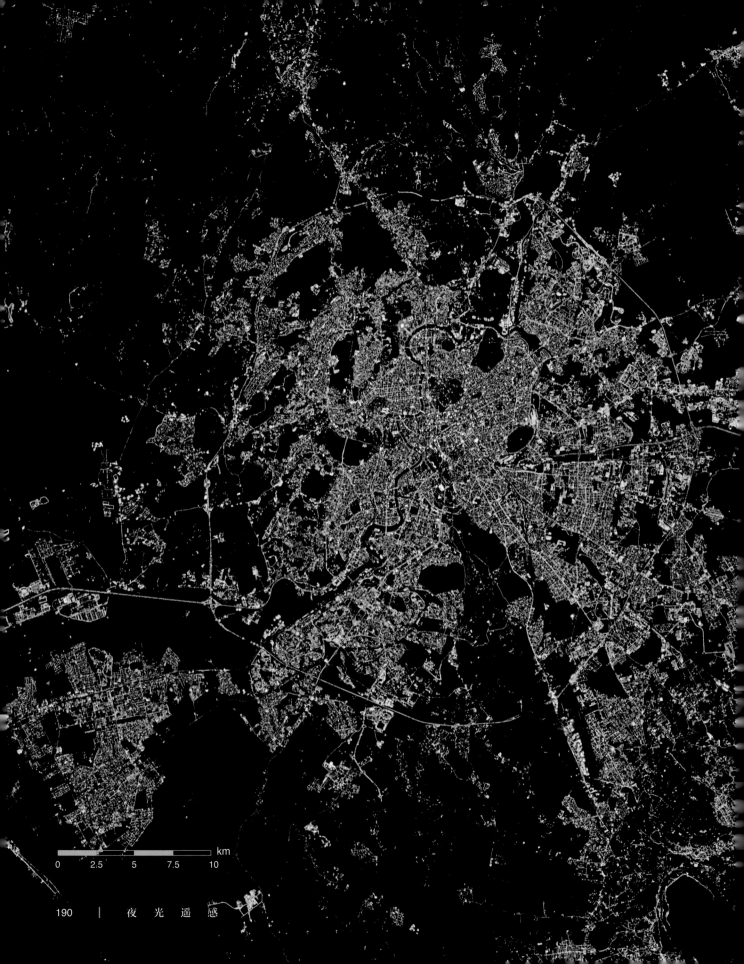

0   2.5   5   7.5   10   km

意大利地图上标注：意 大 利 ◎罗马

意大利 罗马
Rome, Italy

罗马是意大利的首都，也是古罗马帝国的中心，城市面积为 1290 km²，位于意大利半岛的中西部。罗马建在 7 座山丘之上，故被称为"七丘城"，又因建城历史悠久而被称为"永恒之城"。罗马被公认为西方文明和基督教文化的摇篮，并且是天主教会的总部所在地。作为文化、艺术和历史的重要中心，罗马拥有许多著名的古迹、建筑和博物馆。此外，罗马也是现代意大利的政治、经济和文化中心，吸引着来自世界各地的游客和居民。

Rome is the capital of Italy and also the center of the ancient Roman Empire. The city covers an area of 1290 km² and is located in the central-western part of the Italian peninsula. Due to its location atop seven hills, it is referred to as the City of Seven Hills, and due to its ancient history, it is also known as the Eternal City. Rome is recognized as the cradle of Western civilization and Christian culture and is also the headquarters of the Catholic Church. As an important center of culture, art and history, Rome has many famous monuments, buildings and museums. In addition, Rome is also the modern political, economic and cultural center of Italy, attracting tourists and residents from all over the world.

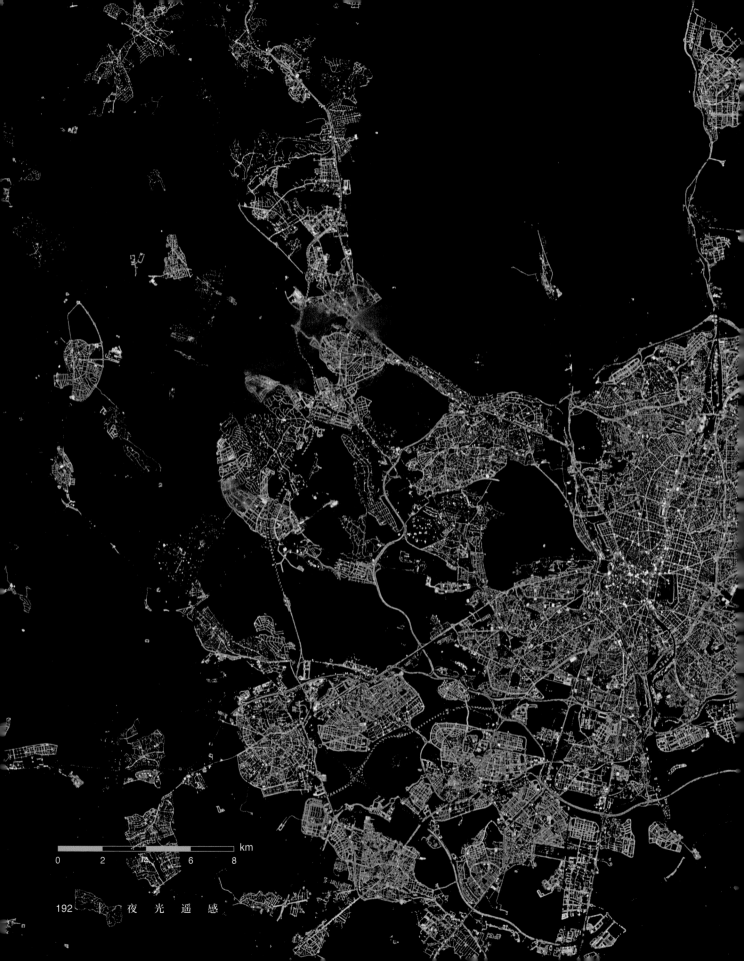

0   2   4   6   8 km

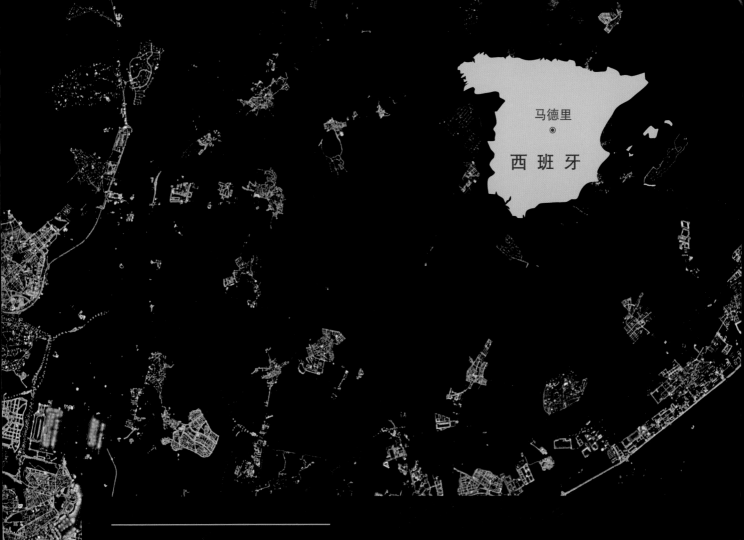

西班牙 马德里

Madrid, Spain

马德里位于伊比利亚半岛中部的曼萨纳雷斯河畔，是西班牙的首都和人口最多的城市，也是西班牙的政治、经济和文化中心。这座城市位于一个较高海拔的平原上，距离最近的海边约 300 km，气候特点是夏热冬凉。作为西班牙最重要的城市之一，马德里在政治、经济和文化领域都发挥着重要作用。马德里拥有丰富的历史、艺术和文化遗产，包括博物馆、剧院、宫殿等，吸引着来自世界各地的游客。马德里也是国际性的商业和金融中心，拥有许多国际公司和机构的总部。

Madrid is located on the banks of the Manzanares River in the central part of the Iberian Peninsula. It is the capital and most populous city of Spain, as well as the country's political, economic and cultural center. The city is situated on a relatively high-altitude plateau, about 300 km away from the nearest coast, with a climate characterized by hot summers and cold winters. As one of Spain's most important cities, Madrid plays an important role in politics, economy and culture. It has a rich historical, artistic and cultural heritage, including museums, theatres, palaces and more, attracting visitors from all over the world. Madrid is also an international business and financial center, with headquarters of many international companies and institutions.

葡萄牙 里斯本
Lisbon, Portugal

　　里斯本是葡萄牙的首都，同时也是该国最大的城市。里斯本位于欧洲大陆最西端，也是大西洋沿岸唯一的首都。里斯本坐落于伊比利亚半岛的西部，濒临大西洋和塔古斯河，占地面积约为 100 km^2。里斯本是葡萄牙的文化、政治和经济中心，拥有着丰富的历史和文化遗产。这座城市充满了古老的街道、古迹、博物馆和艺术品。同时里斯本也是一个充满活力的城市，有着现代化的建筑、购物中心、餐厅和娱乐场所。作为一个沿海城市，里斯本还享有美丽的海滩和壮观的海洋风光。这些因素使得里斯本成为欧洲和世界上备受欢迎的旅游目的地之一。

葡萄牙

⊙里斯本

Lisbon is the capital and largest city of Portugal. It is located at the westernmost end of Europe, and is the only capital on the Atlantic coast. Lisbon lies on the western part of the Iberian Peninsula, overlooking the Atlantic Ocean and the Tagus River. It covers an area of about 100 km^2. Lisbon is Portugal's cultural, political and economic center, with a rich history and cultural heritage. The city is full of ancient streets, monuments, museums and artworks. Lisbon is also a vibrant city with modern buildings, shopping centers, restaurants and entertainment venues. As a coastal city, Lisbon also enjoys beautiful beaches and spectacular ocean views. These factors make Lisbon one of Europe's and the world's most popular tourist destinations.

Km
0   1.25   2.5   3.75   5

瑞

典

斯德哥尔摩 ◉

0 1 2 3 4 km

瑞典 斯德哥尔摩

Stockholm, Sweden

斯德哥尔摩是瑞典的首都和最大的城市，面积约为 187 km²。这座城市的最大特点在于它的城区分散在多个岛屿上，这些岛屿之间通过桥梁连接。斯德哥尔摩是一个历史悠久且充满活力的城市，拥有丰富的文化和历史遗产。作为瑞典的政治、经济和文化中心，斯德哥尔摩是国内外重要的商业和金融中心，也是许多国际公司和机构的总部所在地。该城市还以其高度发达的工业领域而闻名，涵盖金属工业、机械制造、造纸和化工等领域。斯德哥尔摩也是可持续发展和创新的重要中心，致力于推动环保、绿色技术和可持续城市发展。

Stockholm is Sweden's capital and largest city, covering an area of about 187 km². The most distinctive feature of this city is that its urban area is scattered across multiple islands that are connected by bridges. Stockholm is a historic and dynamic city with a rich cultural and historical heritage. As Sweden's political, economic and cultural center, Stockholm is an important domestic and international business and financial center with headquarters of many international companies and institutions. The city is also known for its highly developed industrial sectors covering metal industry, machinery manufacturing, paper making and chemical industry. Stockholm is also an important center for sustainable development and innovation, committed to promoting environmental protection, green technology and sustainable urban development.

芬兰 赫尔辛基
Helsinki, Finland

芬兰

赫尔辛基
◎

km
0   1.25   2.5   3.75   5

　　赫尔辛基面积约为 213 km²，位于芬兰南部，毗邻芬兰湾。作为芬兰的首都，赫尔辛基也是芬兰人口最多的城市。赫尔辛基是芬兰的文化和艺术中心，拥有许多博物馆、剧院、音乐会和艺术展览。同时也是芬兰的经济中心，食品、金属、化工和电子装备生产等是其支柱产业。赫尔辛基的港口也是芬兰的重要国际贸易枢纽，与世界各地保持着紧密地联系。作为一个现代化城市，赫尔辛基致力于可持续发展，投资于生态、社会和经济福祉，努力打造碳中和智慧城市，提高公共服务的效率和居民的幸福感。

　　Helsinki covers an area of about 213 km² and is located in southern Finland, adjacent to the Gulf of Finland. As the capital of Finland, it is also the most populous city in the country. Helsinki is Finland's cultural and artistic center, with many museums, theatres, concerts and art exhibitions. It is also Finland's economic center, with food, metal, chemical and electronic equipment production as its pillar industries. Helsinki's port is also an important international trade hub for Finland, maintaining close ties with the rest of the world. As a modern city, Helsinki is committed to sustainable development, investing in ecological, social and economic well-being, striving to create a carbon-neutral smart city, and improving the efficiency of public services and the happiness of residents.

爱沙尼亚 塔林
Tallinn, Estonia

　　塔林是爱沙尼亚的首都，占地面积 159.4 km²。该城市位于爱沙尼亚西北部的一个海湾上，靠近波罗的海芬兰湾岸边。塔林是爱沙尼亚主要的金融、工业和文化中心。塔林在地理位置上紧邻芬兰塔尔图，距离西北约 186 km，距离芬兰赫尔辛基以南 82 km。塔林的重要性不仅体现在地理位置上，还体现在其丰富的历史和文化遗产上。塔林的老城区被认为是欧洲保存最完好的中世纪城市之一，因此被联合国教科文组织列为世界历史文化遗产。该城市的工业领域覆盖非常广泛，主要涵盖造船与机械制造等领域。此外，塔林在推广数字政府和智慧城市方面有着先进的经验，获得了国际认可。

　　Tallinn is Estonia's capital, covering an area of 159.4 km². The city is located on a bay in northern Estonia, near the coast of the Gulf of Finland in the Baltic Sea. Tallinn is Estonia's main financial, industrial and cultural center. Tallinn is geographically close to Tampere, which is 186 km northwest, and 82 km south of Finland's capital Helsinki. Tallinn's importance is not only reflected in its geographical location, but also in its rich historical and cultural heritage. Tallinn's old town is considered one of the best preserved medieval cities in Europe and is therefore listed as a World Heritage Site by UNESCO. The city's industrial sectors cover a very wide range, mainly including shipbuilding and machinery manufacturing. In addition, Tallinn has advanced experience in promoting digital government and smart city, which has gained international recognition.

拉脱维亚 里加
Riga, Latvia

里加是拉脱维亚的首都和最大的城市，面积约为 302.8 km²。这座城市坐落于里加湾，道加瓦河与波罗的海交汇处。里加的历史可以追溯到 1201 年，曾是汉萨同盟成员。城市的历史中心被联合国教科文组织列为世界文化遗产，以其新青年风格建筑和 19 世纪的木结构建筑而闻名。里加国际机场是波罗的海地区最大、最繁忙的机场，为该地区的重要交通枢纽之一。里加的新艺术风格建筑和文化底蕴吸引着众多游客。作为拉脱维亚的中心，里加在文化、商业和旅游方面发挥着重要的作用，为整个国家的发展做出了重要贡献。

Riga is the capital and largest city of Latvia, covering an area of about 302.8 km². The city is located on the Riga Bay, where the Daugava River meets the Baltic Sea. Riga's history dates back to 1201 and was a member of the Hanseatic League. The city's historic center is listed as a World Heritage Site by UNESCO for its Art Nouveau/Jugendstil architecture and 19th-century wooden buildings. Riga International Airport is the largest and busiest airport in the Baltic region and one of the important transport hubs in the region. Riga's Art Nouveau style architecture and cultural heritage attract many tourists. As the center of Latvia, Riga plays an important role in culture, commerce and tourism, making an important contribution to the development of the whole country.

km

0   1.25   2.5   3.75   5

斯洛伐克

布拉迪斯拉发

斯洛伐克 布拉迪斯拉发
Bratislava, Slovakia

布拉迪斯拉发是斯洛伐克的首都和最大城市，位于多瑙河畔，城市面积约为 368 km²。它与奥地利和匈牙利接壤，因此是唯一一个与两个主权国家接壤的国家首都。多瑙河的毗邻使布拉迪斯拉发成为斯洛伐克的重要交通枢纽，也为城市的发展提供了便利。作为国家首都，布拉迪斯拉发在政治、经济、文化和教育等方面扮演着重要角色。城市的历史和建筑遗产吸引了许多游客，特别是其古老的城市中心。布拉迪斯拉发也是斯洛伐克经济上的重要区域，拥有繁荣的商业和金融活动。

Bratislava is Slovakia's capital and largest city, located on the Danube River and covers an area of about 368 km². It borders Austria and Hungary, making it the only national capital that borders two sovereign countries. The proximity of the Danube River makes Bratislava an important transport hub for Slovakia and also provides convenience for the city's development. As the national capital, Bratislava plays an important role in politics, economy, culture and education. The city's history and architectural heritage attract many tourists, especially its old town area. Bratislava is also an important economic region in Slovakia, with prosperous business and financial activities.

夜 光 遥 感

波兰 华沙
Warsaw, Poland

华沙是波兰的首都和最大的城市，位于波兰中东部的维斯瓦河畔，城市占地 517.24 km²。这座大都市在文化、政治和经济领域具有重要影响力，是波兰的文化中心，拥有丰富的历史和艺术遗产，也是波兰的政治中心，许多国家机构、政府部门和外交机构都设在这里。华沙是波兰重要的经济枢纽，城市内有不少高科技企业、研发中心和创新机构，涵盖了业务流程外包和信息技术外包等领域。华沙在建筑方面与法兰克福和巴黎齐名，它是欧盟摩天大楼数量最多的城市之一，城市风景线充满了现代化的高楼大厦。

Warsaw is Poland's capital and largest city, located on the banks of the Vistula River in central-eastern Poland and covers an area of 517.24 km². This metropolis has a significant influence in culture, politics and economy. Warsaw is Poland's cultural center with a rich historical and artistic heritage. Warsaw is also Poland's political center with many national institutions, government departments and diplomatic agencies located here. Warsaw is an important economic hub in Poland with many high-tech enterprises, research centers and innovation institutions covering fields such as business process outsourcing and information technology outsourcing. Warsaw is comparable to Frankfurt and Paris in terms of architecture, and is one of the cities with the most skyscrapers in the European Union with a modern skyline full of high-rise buildings.

华沙

波 兰

0　1.25　2.5　3.75　5 km

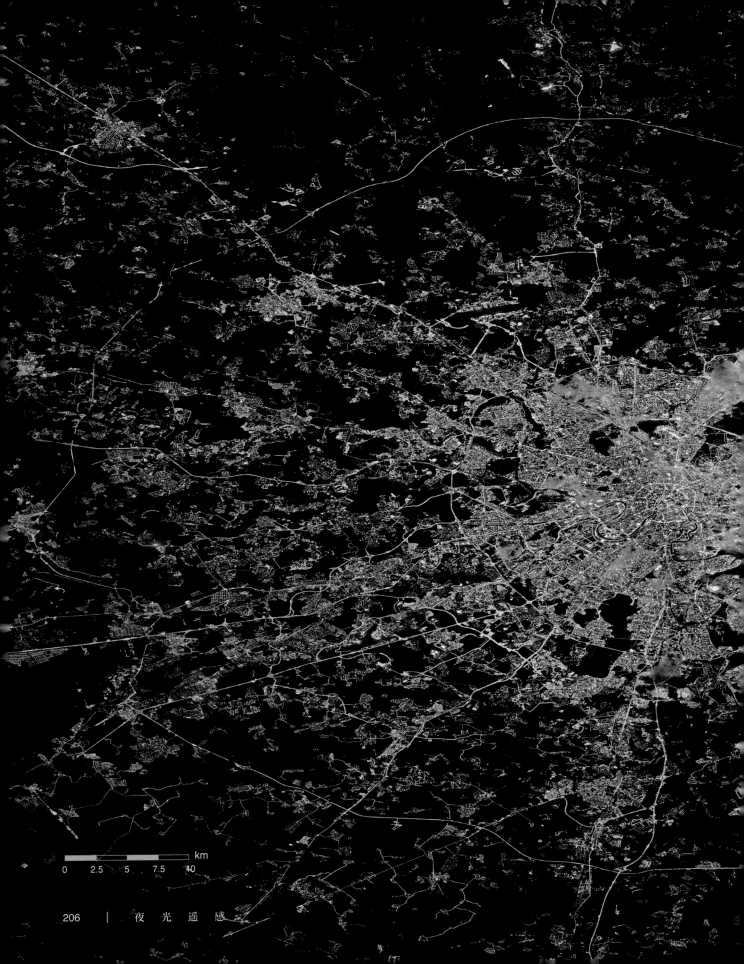

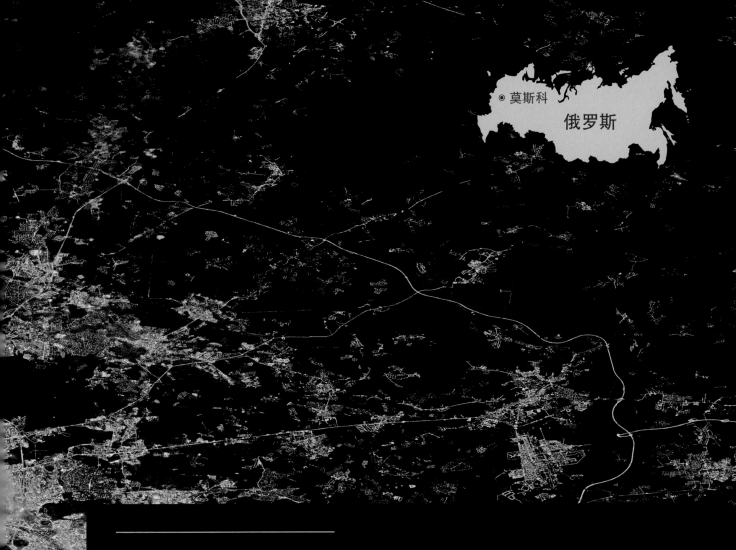

俄罗斯 莫斯科
Moscow, Russia

　　莫斯科位于俄罗斯西部，是俄罗斯的首都和最大的城市，也是俄罗斯和东欧重要的政治、经济、文化、科学和工业中心。截至 2023 年，预估莫斯科的常住人口将达到约 1230 万，城市面积为 2560 km^2。莫斯科在俄罗斯的历史和发展中扮演着重要角色，克林姆林宫作为城市的核心地标，象征着政治和历史的重要性。莫斯科不仅有着悠久的历史，还是俄罗斯文化的中心，拥有许多博物馆、剧院和艺术展览，吸引了来自世界各地的游客。且该城市规划和交通系统也相当发达，多个环线的地铁系统和现代化的交通网络使城市内外的交通更加便捷。

Moscow is located in western Russia and is the capital and largest city of the country. It is also an important political, economic, cultural, scientific and industrial center of Russia and Eastern Europe. As of 2023, Moscow's resident population is estimated to reach about 12.3 million, with a city area of 2560 km^2. Moscow plays an important role in Russia's history and development. The Kremlin, as the core landmark of the city, symbolizes its political and historical significance. Moscow not only has a long history, but also is the center of Russian culture, with many museums, theatres and art exhibitions that attract visitors from all over the world. The city also has a developed urban planning and transportation system, with multiple ring subway systems and modern transportation networks that facilitate traffic within and outside the city.

白俄罗斯 明斯克
Minsk, Belarus

　　明斯克位于白俄罗斯的中部，是该国的首都和最大城市，面积为409.5 km²。明斯克在白俄罗斯的历史和发展中具有重要地位，这座城市以其多样的文化活动、历史古迹和现代化设施而闻名。明斯克的工业部门集中在多个领域，其中包括拖拉机制造、电机马达生产、家用电器和纺织等。明斯克还拥有丰富的文化生活，包括剧院、博物馆、音乐会和艺术展览。城市内的历史古迹和建筑也吸引了许多游客前来参观。

　　Minsk is located in the central part of Belarus and is the capital and largest city of the country. It covers an area of 409.5 km². The city has an important position in Belarus' history and development. It is famous for its diverse cultural activities, historical monuments and modern facilities. Minsk's industrial sectors are concentrated in various fields, including tractor manufacturing, electric motor production, household appliances and textiles. Minsk also has a rich cultural life, including theatres, museums, concerts and art exhibitions. The historical and architectural attractions in the city also attract many tourists to visit.

0　　　　1.25　　　　2.5　　　　3.75　　　　5　km

Vienna is the capital of Austria and also the largest city on the Danube River. It covers an area of 414.65 km^2. It is not only one of Austria's nine federal states, but also the country's cultural, economic and political center. Vienna is Austria's most populous city. Vienna is known as the "City of Music" due to its rich musical heritage. Many renowned classical musicians have composed exceptional works in this city, leaving behind stories of their work and lives, which have contributed to Vienna's cultural richness and portrait of the city. Vienna's urban landscape and historical landmarks attract many tourists to visit, including classical buildings, museums, art galleries and more. It is also praised for its rich musical heritage and beautiful scenery.

km
0   1.25   2.5   3.75   5

维也纳 ◉

奥 地 利

奥地利 维也纳
Vienna, Austria

　　维也纳是奥地利的首都，面积为 414.65 km^2。它不仅是奥地利的九个联邦州之一，还是奥地利的文化、经济和政治中心。维也纳是奥地利人口最多的城市。维也纳因其丰富的音乐遗产而被称为"音乐之城"，许多著名的古典音乐家曾在此创作出优秀的作品，留下许多工作和生活的故事，为维也纳增添了文化底蕴和城市特色。维也纳的城市景观和历史地标吸引了许多游客前来参观，包括古典建筑、博物馆、艺术画廊等，同时它也因其丰富的音乐遗产和美丽的景观而备受赞誉。

基辅

乌克兰

乌克兰 基辅
Kiev, Ukraine

　　基辅位于乌克兰中北部的第聂伯河两岸，城区面积达 827 km²，是乌克兰首都和人口最多的城市之一。作为乌克兰的首都，基辅在东欧地区具有重要的地位。这座城市不仅是政治和行政中心，还是乌克兰的工业、科学、教育和文化中心。基辅拥有广泛的公共交通网络，以便于市民和游客在城市内进行出行。基辅拥有机械生产、航空制造等产业，这些工业区域主要集中于城市西部。基辅也是一个充满历史和文化遗产的城市。许多古老的建筑、教堂、博物馆和艺术画廊都见证了这座城市丰富的历史。基辅还以其大学和科研机构的存在而闻名，为乌克兰的教育和科学发展做出了贡献。

　　Kyiv is located on both banks of the Dnieper River in north-central Ukraine. The city covers an area of 827 km² and is the capital and one of the most populous cities of Ukraine. As the capital of Ukraine, Kyiv has an important position in Eastern Europe. The city is not only a political and administrative center, but also Ukraine's industrial, scientific, educational and cultural center. Kyiv has a developed public transport network to facilitate travel within the city for residents and tourists. Kyiv's industrial sectors include machinery production, aviation manufacturing and more. These industrial areas are mainly concentrated in the western part of the city. Kyiv is also a city full of historical and cultural heritage. Many ancient buildings, churches, museums and art galleries reflect the rich history of this city. Kyiv is also known for its universities and research institutions that contribute to Ukraine's education and scientific development.

匈牙利 布达佩斯
Budapest, Hungary

　　布达佩斯是匈牙利的首都和人口最多的城市，城市面积为 525.2 km²。布达佩斯拥有发达的公共交通网络，城市居民大都选择公共交通工具来减少交通拥堵和环境污染。布达佩斯的经济呈现多元化的发展趋势，第二产业和第三产业都在持续增长。布达佩斯也是匈牙利文化、艺术和历史的中心。这座城市拥有许多博物馆、历史古迹、音乐和艺术场所，吸引了许多游客前来参观和探索。

　　Budapest is Hungary's capital and most populous city. The city covers an area of 525.2 km². Budapest has a developed public transport network that allows most residents to choose public transport to reduce traffic congestion and environmental pollution. Budapest's economy shows a diversified development trend, with both secondary and tertiary industries growing steadily. Budapest is also Hungary's cultural, artistic and historical center. The city has many museums, historical monuments, music and art venues that attract many visitors to explore.

◎布达佩斯

匈 牙 利

摩尔多瓦 基希讷乌
Chisinau, Moldova

摩尔多瓦

基希讷乌

km
0   1   2   3   4

　　基希讷乌是摩尔多瓦共和国的首都和最大城市，位于该国的中部地区，城区面积为 565 km²。该城市不仅是摩尔多瓦经济最繁荣的地区，也是摩尔多瓦最大的交通枢纽。基希讷乌是一座白色的城市，在城市的大街小巷里，到处可以看见掩映在绿荫丛中的白色建筑，这些白色的建筑都是用摩尔多瓦境内盛产的石灰石为材料建成的。洁白的建筑与翠绿的树木交相辉映，使基希讷乌显示出一派生机与活力。城市街道呈棋盘状，列宁大街和普希金大街是城中主要街道。基希讷乌境内有 36 所大学，以及摩尔多瓦科学院。自从摩尔多瓦独立后，整个城市就逐渐变得较有生气且发展健全，居民生活水准显著提高。

Chisinau is the capital and largest city of Moldova Republic, located in the central region of the country. The city covers an area of 565 km². The city is not only Moldova's most prosperous economic region, but also Moldova's largest transport hub. Chisinau is a white city, where throughout its streets and alleys, white buildings concealed in greenery can be seen. These white structures are constructed using the abundant limestone found within the Moldova region. The pristine buildings harmonize with the lush trees, imparting Chisinau with a sense of vibrancy and vitality. The city's streets form a chessboard pattern, with Lenin Street and Pushkin Street serving as the main thoroughfares. Chisinau boasts 36 universities and the Moldova Academy of Sciences. Since Moldova's independence, the entire city has gradually become more vibrant and well-developed, leading to a significant improvement in the standard of living for its residents.

黑山 波德戈里察
Podgorica, Montenegro

　　波德戈里察是黑山共和国的首都和最大城市，位于巴尔干半岛的中西部，且被山区环绕，城市面积为 1500 km^2。波德戈里察坐落在斯卡达尔湖的北部，靠近亚得里亚海的沿岸。斯卡达尔湖是巴尔干半岛最大的湖泊之一，为波德戈里察增添了自然美景，亚得里亚海则为城市提供了连接外界的海上通道。波德戈里察因独特的自然环境、悠久的历史遗迹、现代化的建筑和悠闲的氛围而独具魅力。

Podgorica is the capital and largest city of Montenegro, located in the northern part of the Balkan Peninsula. It is surrounded by mountains and covers an area of 1500 km^2. Podgorica lies on the northern part of Lake Skadar, near the coast of the Adriatic Sea. Lake Skadar is one of the largest lakes in the Balkan Peninsula, adding natural beauty to Podgorica. The Adriatic Sea provides the city with a maritime connection to the outside world. Podgorica is attractive due to its unique natural environment, rich historical heritage, modern architecture, and laid-back atmosphere.

黑　山

波德戈里察
◎

```
                                          km
0        1.25      2.5      3.75      5
```

罗马尼亚

布加勒斯特
◉

罗马尼亚 布加勒斯特
Bucharest, Romania

　　布加勒斯特是罗马尼亚的首都和最大的城市。它位于罗马尼亚东南部，登博维察河畔，与保加利亚接壤，是罗马尼亚的文化、工业和金融中心，也是欧洲最大最现代化和发展最快的国际大都市之一。布加勒斯特市区12个湖泊同登博维察河相平行，一个连着一个，宛如一串珠光闪闪的项链，把布加勒斯特装扮得分外艳丽。城市北郊有著名的伯尼亚萨森林，市内用草坪、玫瑰花、月季花组成的色彩缤纷的花坛随处可见。植树造林，养花种草，绿化城市，美化环境，已成为布加勒斯特居民的传统和爱好。如今的布加勒斯特，绿荫如盖，花木成林，湖水片片，成为一座花园城市。

Bucharest is the capital and largest city of Romania. It is located in the southeast of Romania, on the banks of the Dâmbovița River, and the border with Bulgaria. It is described as the cultural, industrial and financial center of Romania and it is also one of the largest, most modern and fastest-growing international metropolises in Europe. Within the city of Bucharest, there are 12 lakes that run parallel to the Dâmbovița River, like a string of sparkling pearls, adorning the city with vibrant beauty. The famous Băneasa Forest lies to the north of the city, while colorful flower beds made of lawns, roses, and roses can be found throughout the city. Tree planting, landscaping, and beautifying the environment have become traditions and passions of Bucharest's residents. Present-day Bucharest is covered with greenery, filled with flowers, and dotted with lakes, making it a city resembling a garden.

贝尔格莱德

塞尔维亚

塞尔维亚 贝尔格莱德

Belgrade, Serbia

贝尔格莱德是塞尔维亚的首都和最大的城市。它位于萨瓦河和多瑙河的交会处，是潘诺尼亚平原和巴尔干半岛的十字路口。贝尔格莱德是塞尔维亚中央政府、行政机构和政府部门的所在地，也是规模较大的公司、媒体和科学机构的所在地。贝尔格莱德属于大陆性气候，全年平均气温为 11.7℃，最热的月份在 7 月，平均温度 22.1℃。全年气温 30℃以上有 31 天，25℃以上有 95 天。贝尔格莱德年降雨量为 700 mm。根据 2022 年人口普查，贝尔格莱德大都市区的人口为 168 万人，面积约为 360 km²。

Belgrade is the capital and largest city in Serbia. It is located at the confluence of the Sava and Danube rivers and the crossroads of the Pannonian Plain and the Balkan Peninsula. Belgrade is the seat of Serbia's central government, administration and ministries, as well as larger companies, media and scientific institutions. Belgrade has a continental climate, with an average annual temperature of 11.7℃. The hottest month is July, with an average temperature of 22.1℃. There are 31 days in the year when temperatures exceed 30℃, and 95 days when temperatures exceed 25℃. The city receives an annual rainfall of 700 millimeters. According to the 2022 census, the population of the Belgrade metropolitan area is 168 million and the urban area is about 360 km².

北马其顿 斯科普里
Skopje, Northern Macedonia

　　斯科普里位于瓦尔达尔河上游，是北马其顿的首都和最大的城市，是该国的政治、经济、文化和学术中心，在古罗马时期的名称是斯库皮。斯科普里附近地区自约前 4000 年以来就有人居住，斯科普里市中心的斯科普里城堡附近曾发现新石器时代的集落遗迹。斯科普里市区附近的山地由地震活动而形成，西部有夏尔山，南部是雅库皮察山脉，东部是奥索戈沃。斯科普里市中心以瓦尔达尔河为界，分为两个行政区。河北岸是赛尔地区，斯科普里旧巴扎就位于这里，是斯科普里的旧市区。南岸是辛塔尔，有众多的现代建筑，是斯科普里市区的中心。

　　Skopje is located on the upstream of the Vardar River and is the capital and largest city of North Macedonia. It serves as the political, economic, cultural, and academic center of the country. In ancient Roman times, it was known as Scupi. The area near Skopje has been inhabited for over 4000 years, with traces of Neolithic settlements found near the Skopje Fortress in the city center. The surrounding mountains of Skopje were formed due to seismic activity. The Shar Mountains are located to the west, the Jakupica Mountains to the south, and the Osogovo Mountains to the east. The city center of Skopje is divided by the Vardar River into two administrative districts. The northern bank is called the Šar Region, where the Old Bazaar of Skopje is located, representing the city's historic district. The southern bank is known as Centar, hosting numerous modern buildings and serving as the central area of Skopje's cityscape.

⊙ 斯科普里

北马其顿

```
                                              km
0      1.25      2.5      3.75      5
```

阿尔巴尼亚 地拉那
Tirana, Albania

　　地拉那是阿尔巴尼亚的首都和最大的城市，位于该国中部，周围群山环抱，西北方有一个小山谷，可以俯瞰远处的亚得里亚海。国家中心的重要地理位置及拥有现代化的空中、海上、铁路和公路交通，使得地拉那成为阿尔巴尼亚重要的经济、金融、政治和贸易中心。地拉那是一座历史悠久的都城，早在 2700 多年前，阿尔巴尼亚人的祖先就在这里生活。如今，地拉那人口近 93 万。从飞机上俯瞰，可见城市的北、东、南三面为山丘环抱，山上长满绿色的地中海植物。

Tirana is the capital and largest city of Albania. It is located in the center of the country, enclosed by mountains and hills and a slight valley to the northwest overlooking the Adriatic Sea in the distance. Due to the important geographical location of the national center and the modern air, sea, railway and road transportation, Tirana is an important economic, financial, political and trade center of Albania. Tirana is an ancient capital city with a history dating back over 2700 years, where the ancestors of Albanians lived. Today, Tirana has a population of nearly 0.93 million. When viewed from above, the city is surrounded by hills to the north, east, and south, covered in lush Mediterranean vegetation.

◎地拉那

阿尔巴尼亚

km
0　1　2　3　4

希 腊

◎ 雅典

希腊 雅典
Athens, Greece

雅典是希腊的首都和第一大城市，市区人口超过三百万，是欧盟的第八大城市，也是希腊经济、政治、工业和文化中心。雅典是世界上最古老的城市之一，其有记载的历史长达 3400 多年。因其对古罗马的文化和政治产生了深远影响，被称为西方文明的摇篮和民主的发源地。2022 年雅典人口为 379.2 万人，城市面积约为 412 km^2。

Athens is the capital and largest city of Greece. With its urban area's population numbering over three million, it is the eighth largest urban area in the European Union. Athens is one of the world's oldest cities, with its recorded history spanning over 3400 years. It is widely referred to as the cradle of Western civilization and the birthplace of democracy, largely because of its cultural and political influence on Ancient Rome. In 2022, the population of Athens is 3.792, and the urban area is about 412 km^2.

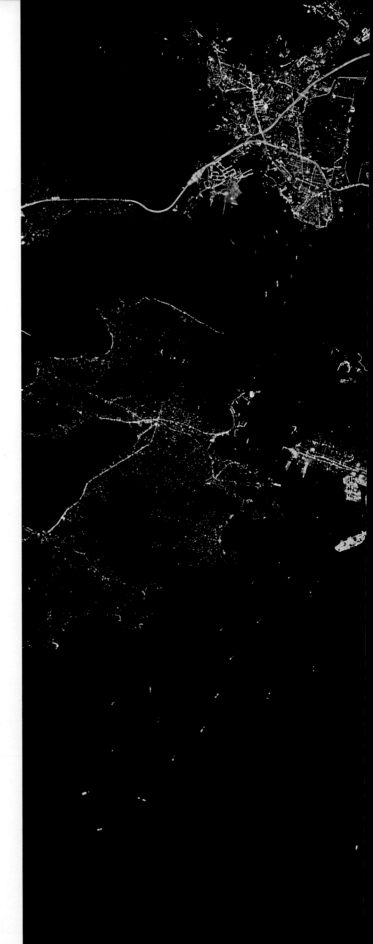

km
0   1.25   2.5   3.75   5

波斯尼亚和黑塞哥维那
萨拉热窝
Sarajevo, Bosnia and Herzegovina

波斯尼亚和黑塞哥维那

萨拉热窝 ◎

　　萨拉热窝是波黑的首都和第一大城市，面积约为 142 km²。它是波黑的政治、金融、社会和文化中心，也是巴尔干半岛的重要文化中心，在娱乐、时尚和艺术领域具有区域性影响力。由于其悠久的宗教和文化多样性历史，萨拉热窝有时被称为"欧洲的耶路撒冷"或"巴尔干的耶路撒冷"。

　　Sarajevo is the capital and largest city of Bosnia and Herzegovina, covering an area of approximately 142 km². Sarajevo is the political, financial, social and cultural center of Bosnia and Herzegovina and a prominent center of culture in the Balkans. It exerts region-wide influence in entertainment, media, fashion and the arts. Due to its long history of religious and cultural diversity, Sarajevo is sometimes called the "Jerusalem of Europe" or "Jerusalem of the Balkans".

| 0 | 1.25 | 2.5 | 3.75 | 5 | km |

塞浦路斯 尼科西亚
Nicosia, Cyprus

　　尼科西亚是塞浦路斯的首都和第一大城市，面积约为 50.5 km²，位于梅索里亚平原中部，派迪亚斯河河畔，北依横跨岛国北岸的凯里尼亚山脉，西南同特罗多斯山遥遥相望。尼科西亚是所有欧盟成员国首都中最东南的一个，是塞浦路斯的政治、经济和文化中心。2021 年尼科西亚人口为 35.2 万人。

Nicosia is the capital and largest city of The Republic of Cyprus, covering an area of approximately 50.5 km². It is located near the center of the Mesaoria plain, on the banks of the River Pedieos, and is bounded to the north by the Kyrenia Mountains, which straddle the northern coast of the island, and to the southwest by the Troodos Mountains. Nicosia is the southeastern most of all European Union member states' capitals and the political, economic and cultural center of Cyprus. In 2021, the population of Nicosia is about 0.352 million.

尼科西亚

塞浦路斯

克罗地亚 萨格勒布
Zagreb, Croatia

　　萨格勒布是克罗地亚的首都和第一大城市，位于该国西北部，萨瓦河沿岸，梅德韦德尼察山南坡，靠近克罗地亚和斯洛文尼亚之间的国际边界，海拔约 122 m。萨格勒布是政府行政机构所在地，此外克罗地亚几乎所有的大公司、媒体和科研机构的总部也都设在萨格勒布。中欧、地中海和东南欧在此交汇，使萨格勒布地区成为克罗地亚公路、铁路和航空网络的中心。2022 年萨格勒布人口为 80 万人，城市面积约为 1291 km²。

Zagreb is the capital and largest city of Croatia. It is in the northwest of the country, along the Sava river, at the southern slopes of the Medvednica mountain. Zagreb stands near the international border between Croatia and Slovenia at an elevation of approximately 122 m above sea level. Zagreb is the seat of the central government, administrative bodies, and almost all government ministries. Almost all of the largest Croatian companies, media, and scientific institutions have their headquarters in the city. Here Central Europe, the Mediterranean and Southeast Europe meet, making the Zagreb area the center of the road, rail and air networks of Croatia. In 2022, the population of Zagreb is 0.8 million and the urban area is about 1291 km².

格勒布
⊙
克罗地亚

AMERICA 美洲

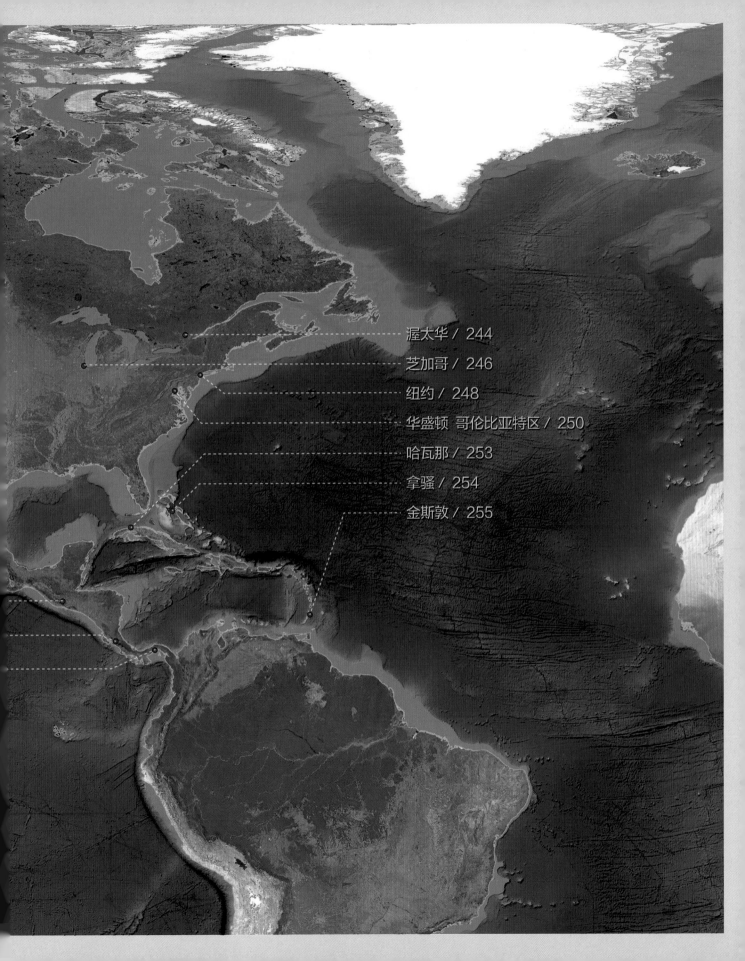

加拿大 温哥华
Vancouver, Canada

　　温哥华是加拿大西南部的主要城市，是加拿大第三大城市，全球最宜居的城市之一，位于不列颠哥伦比亚省的低陆平原地区。2021年大温哥华地区的人口约为264万，是加拿大人口密度最大的城市，每平方公里超过5700人。温哥华港是加拿大最繁忙和最大的港口，也是北美最多元化的港口。温哥华森林广袤，吸引了大量游客，因此其旅游业也成为了主要产业之一，每年有超过一百万人乘坐游轮前往温哥华度假。

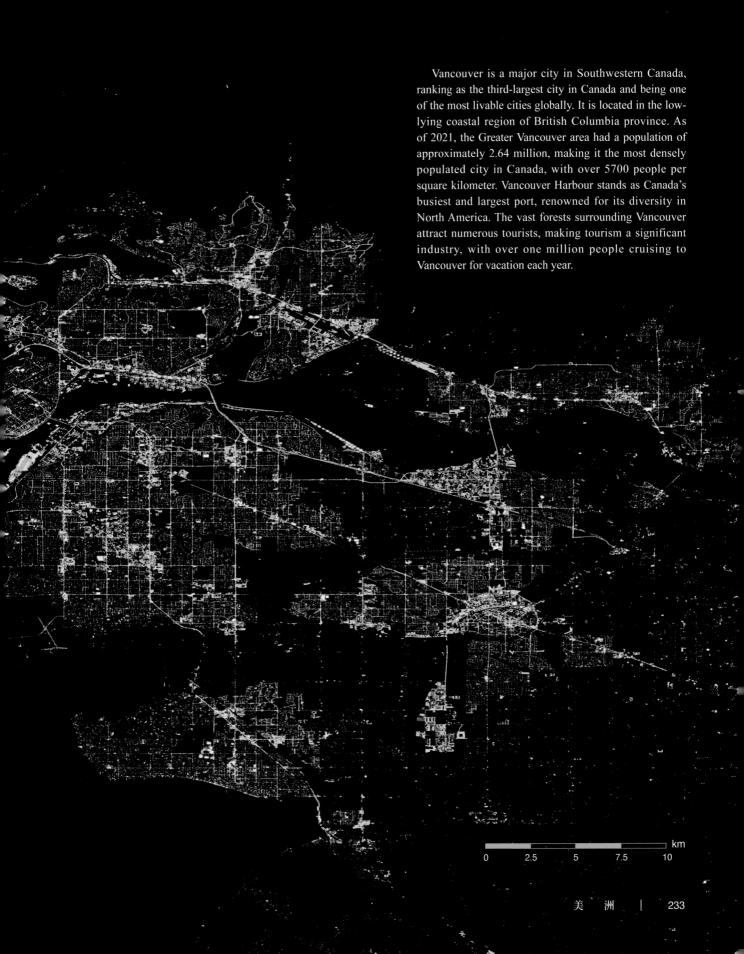

Vancouver is a major city in Southwestern Canada, ranking as the third-largest city in Canada and being one of the most livable cities globally. It is located in the low-lying coastal region of British Columbia province. As of 2021, the Greater Vancouver area had a population of approximately 2.64 million, making it the most densely populated city in Canada, with over 5700 people per square kilometer. Vancouver Harbour stands as Canada's busiest and largest port, renowned for its diversity in North America. The vast forests surrounding Vancouver attract numerous tourists, making tourism a significant industry, with over one million people cruising to Vancouver for vacation each year.

km
0   2.5   5   7.5   10

km
0   2.5   5   7.5   10

美国 洛杉矶
Los Angeles, United States

　　洛杉矶是美国南加州的商业、金融和文化中心。它是加利福尼亚州最大的城市，仅次于纽约市，是美国人口第二多的城市。截至 2021 年，洛杉矶市区人口约为 390 万，以其地中海气候、种族和文化多样性而闻名，同时洛杉矶还是好莱坞电影业的发源地，拥有庞大的都市区。该市大部分位于南加州的盆地，西邻太平洋，穿过圣莫尼卡山脉，向北延伸至圣费尔南多谷，东部则与圣盖博谷接壤。洛杉矶拥有多元化的经济结构和广泛的工业基础，是洛杉矶港美洲最繁忙的集装箱港口之一。

　　Los Angeles is the commercial, financial, and cultural center of Southern California, United States. It is the largest city in California, the second-most populous city in the United States after New York City. With a population of roughly 3.9 million residents within the city limits as of 2021, Los Angeles is known for its Mediterranean climate, ethnic and cultural diversity, being the home of the Hollywood film industry, and its sprawling metropolitan area. The majority of the city proper lies in a basin in Southern California adjacent to the Pacific Ocean in the west and extending partly through the Santa Monica Mountains and north into the San Fernando Valley, with the city bordering the San Gabriel Valley to its east. Los Angeles has a diverse economy with a broad range of industries. The Ports of Los Angeles is one of the busiest container ports in the Americas.

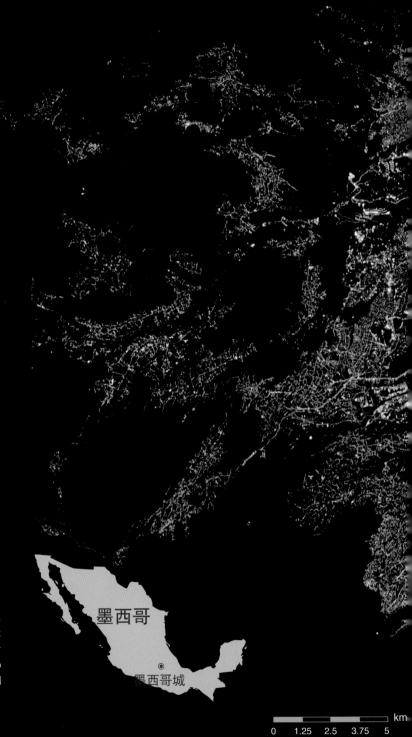

墨西哥 墨西哥城
Mexico City, Mexico

　　墨西哥城是墨西哥的首都和第一大城市，也是北美洲人口最多的城市。该城面积 1525 km²，人口约 2280 万（含卫星城），海拔 2240 m。墨西哥的首都既是整个美洲最古老的首都，又是古印第安人建立的两个城市之一。公元 1325 年左右，墨西卡人（阿兹特克人）在特克斯科科湖的一群岛屿上建造了这座城市的雏形，并命名为特诺奇蒂特兰。

　　Mexico City is the capital and largest city of Mexico, and the most populous city in North America. The city has an area of 1525 km², a population of about 22.8 million (including Satellite Cities), and an elevation of 2,240m above sea level. Mexico's capital is also both the oldest capital city in the Americas and one of two founded by the Paleo-Indians. The city was originally built on a group of islands in Lake Texcoco by the Mexica (Aztecs) around 1325 A. D., under the name Tenochtitlan.

墨西哥

墨西哥城

km
0   1.25   2.5   3.75   5

危地马拉

◉ 危地马拉城

危地马拉 危地马拉
Guatemala, Guatemala

　　危地马拉是危地马拉的首都和第一大城市，也是美洲大陆中部地区人口最多的城市。该市位于国家中南部，是危地马拉的政治、文化和经济中心。该市还是危地马拉的主要交通枢纽，拥有拉奥罗拉国际机场，并且是危地马拉大部分高速公路干道的起点或终点。2022 年危地马拉人口为 301.5 万人，城市面积约为 996 km²，海拔 1480 m。

Guatemala is the capital and largest city of Guatemala, and the most populous urban area in the central part of the American continent. Located in the south-central part of the country, the city is the political, cultural and economic center of Guatemala. The city is also Guatemala's main transportation hub, hosting La Aurora International Airport, and is the start or end point of most of Guatemala's arterial highways. In 2022, the population of Guatemala is 3.015 million, and the city area is about 996 km² with an elevation of 1480 m above sea level.

哥斯达黎加

圣何塞

0 1.25 2.5 3.75 5 km

哥斯达黎加 圣何塞

San José, Costa Rica

　　圣何塞是哥斯达黎加的首都和第一大城市，也是哥斯达黎加省的首府。它位于国家的中心位置，中央山谷中西部，圣何塞州境内。圣何塞是哥斯达黎加的国家政府所在地，政治和经济中心，以及重要的交通枢纽。圣何塞在拉丁美洲城市中以其高质量的生活水平、安全性、全球化水平、环境表现和良好的公共服务而闻名。根据对拉丁美洲的研究，圣何塞是当地最安全、暴力事件最少的城市之一。圣何塞城市人口为 163 万人，城市面积约为 4966 km^2。

　　San José is the capital and largest city of Costa Rica, and the capital of the province of the same name. It is in the center of the country, in the mid-west of the Central Valley, within San José Canton. San José is Costa Rica's seat of national government, focal point of political and economic activity, and major transportation hub. San José is notable among Latin American cities for its high quality of life, security, level of globalization, environmental performance and public service. According to studies on Latin America, San José is one of the safest and least violent cities in the region. The population of San José is 1.63 million, and the urban area is approximately 4966 km^2.

巴拿马 巴拿马城
Panama City, Panama

巴拿马城，是巴拿马的首都和第一大城市。该市是国家的政治和经济中心，也是文化、金融中心，位于巴拿马省巴拿马运河的太平洋入口处，是美洲大陆最重要的贸易路线之一的中转站，通往诺姆布雷德迪奥斯和波托贝洛集市，西班牙从美洲开采的大部分矿物都经过这里。巴拿马城人口为 88 万人，城市面积约为 275 km²。

Panama City is the capital and largest city of Panama. The city is the political and economic center of the country, as well as a hub for culture and finance. The city is located at the Pacific entrance of the Panama Canal, in the province of Panama. It was a stopover point on one of the most important trade routes in the American continent, leading to the fairs of Nombre de Dios and Portobelo, through which passed most of the minerals that Spain mined from the Americas. The population of Panama City is 0.88 million, and the city area is about 275 km².

巴拿马城

巴拿马

加拿大

渥太华

加拿大 渥太华

Ottawa, Canada

渥太华是加拿大的首都，位于安大略省南部，渥太华河和里多运河的交会处。渥太华与魁北克省的加蒂诺接壤，为渥太华－加蒂诺人口普查都市区和首都圈的核心区域。渥太华是加拿大的政治中心和联邦政府总部，许多外国大使馆、重要建筑、组织机构和加拿大政府坐落于此。首都地区（包括安大略省渥太华市、魁北克省加蒂诺市及周围城镇）人口约为 132.4 万人，面积约为 4715 km²。

Ottawa the capital city of Canada is located at the confluence of the Ottawa River and the Rideau River in the southern portion of the province of Ontario. Ottawa borders Gatineau, Quebec, and forms the core of the Ottawa–Gatineau census metropolitan area and the National Capital Region. Ottawa is the political center of Canada and headquarters to the federal government and houses numerous foreign embassies, key buildings, organizations, and institutions of Canada's government. The Capital Region (including Ottawa, Ontario, Gatineau, Quebec, and surrounding towns) has a population of approximately 1.324 million and an area of approximately 4715 km².

美国 芝加哥
Chicago, United States

　　芝加哥是美国伊利诺伊州人口最多的城市，也是美国人口第三多的城市，仅次于纽约和洛杉矶。芝加哥还是金融、文化、商业、工业、教育、科技、电信和交通的国际中心。该地区是美国的铁路枢纽，拥有最多的联邦高速公路。芝加哥的奥黑尔国际机场是世界上最繁忙的六大机场之一。2022 年芝加哥人口约为 267 万人，城市面积约为 606.2 km^2。

　　Chicago is the most populous city in the U.S. state of Illinois and the third-most populous in the United States after New York City and Los Angeles. Chicago is an international hub for finance, culture, commerce, industry, education, technology, telecommunications, and transportation. The region is the nation's railroad hub and has the largest number of federal highways. O'Hare International Airport in Chicago is routinely ranked among the world's top six busiest airports. In 2022, the population of Chicago is about 2.67 million, and the urban area is about 606.2 km^2.

km
0   2.5   5   7.5   10

芝加哥

美　国

美国 纽约
New York, United States

　　纽约是美国人口最多的城市，截至 2021 年，纽约市人口为 882.35 万人，面积为 1214 km²，是美国人口最稠密的城市，人口数量是美国第二大城市洛杉矶的两倍多。纽约市坐落于世界上最大的天然港口之一，由五个行政区组成，每个行政区都与纽约州的一个县接壤。纽约市位于纽约州的南部，是全球的文化、金融、高新科技和娱乐中心。纽约还是联合国总部所在地，是重要的国际外交中心，被称为世界之都，许多街区和纪念碑都是纽约市的重要地标，如位于百老汇剧院区灯火通明地带的时代广场，是世界上最繁忙的十字路口，也是世界娱乐产业的中心。

　　New York is the most populous city in the United States. With a 2021 population of 8.8235 million distributed over 1214 km², New York City is the most densely populated major city in the United States and more than twice as populous as Los Angeles, the nation's second-largest city. Situated on one of the world's largest natural harbors, New York City comprises five boroughs, each of which is coextensive with a respective county of the state of New York. New York City is located at the southern tip of New York State and is a global cultural, financial, high-tech, entertainment center. Home to the headquarters of the United Nations, New York is an important center for international diplomacy, and is sometimes described as the capital of the world. Many districts and monuments in New York City are major landmarks. Times Square is the brightly illuminated hub of the Broadway Theater District, the world's busiest pedestrian intersections and a major center of the world's entertainment industry.

美 国

© 纽约

km
0   2.5   5   7.5   10

华盛顿

美 国

美国 华盛顿哥伦比亚特区

Washington, D.C., United States

　　华盛顿哥伦比亚特区简称华盛顿特区，是美国的首都和联邦区，位于美国东北部，由美国国会直接管辖，因此不属于美国的任何一州。作为美国联邦政府和多个国际组织的所在地，华盛顿是非常重要的世界政治中心。该城市人口约为 69 万人，城市面积约为 177km²。在气候上，属副热带湿润气候区，四季分明，气温变化相对和缓，全年降水分配均匀。冬季冷凉，微潮；夏季相对炎热潮湿。

　　Washington, D.C. formally the District of Columbia, is the capital city and federal district of the United States. The city is located on the east bank of the Potomac River, which forms its southwestern border with Virginia, and borders Maryland to its north and east. As the seat of the U.S. federal government and several international organizations, the city is an important world political capital. The population of Washington is about 0.69 million, and the urban area is about 177 km². Washington D.C., falls within the subtropical humid climate zone, characterized by distinct seasons, relatively moderate temperature variations, and evenly distributed precipitation throughout the year. Winters are cool and somewhat damp, while summers are relatively hot and humid.

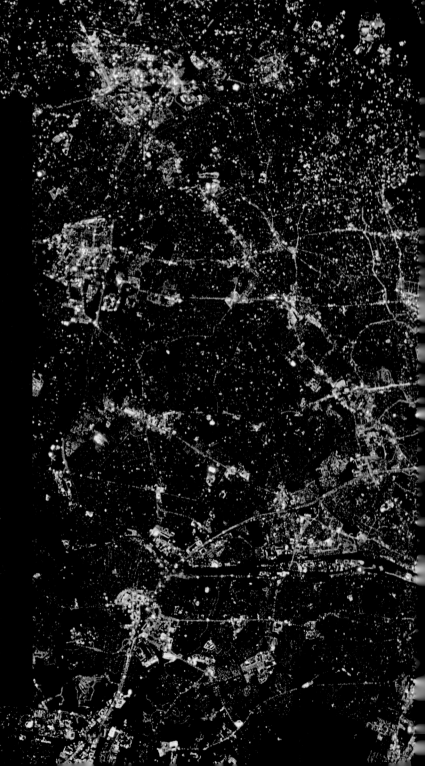

哈瓦那

古
巴

km
0 0.5 1 1.5 2

古巴 哈瓦那
Havana, Cuba

　　哈瓦那是古巴的首都和第一大城市，是典型的热带气候。哈瓦那是哈瓦那省的中心，也是古巴的主要港口和商业中心。哈瓦那的港湾狭长，近岸水深 12 m，可容纳远洋巨轮，湾底建有隧道，沟通两岸交通，全国一半以上的进出口货物经此。该城市也是加勒比海地区最大城市，许多政体机构以及商业总部也都设在哈瓦那。1982 年，哈瓦那老城被联合国教科文组织列为世界文化遗产。这座城市也因其历史、文化、建筑和古迹而闻名。2023 年哈瓦那城市人口约为 214.9 万人，城市面积约为 721km^2。

　　Havana is the capital and largest city of Cuba. As typical of Cuba, Havana experiences a tropical climate. The heart of the La Habana Province, Havana is the country's main port and commercial center. Havana's harbor is long and narrow, with a depth of 12 m near the shore, which can accommodate ocean-going vessels, and there are tunnels at the bottom of the bay to connect the two sides of the river, through which more than half of the country's imports and exports pass. It is also the largest city in the Caribbean, and many political institutions and business headquarters are located in Havana. Old Havana was declared a UNESCO World Heritage Site in 1982. The city is also noted for its history, culture, architecture and monuments. In 2023, the urban population of Havana is about 2.149 million, and the urban area is about 721 km^2.

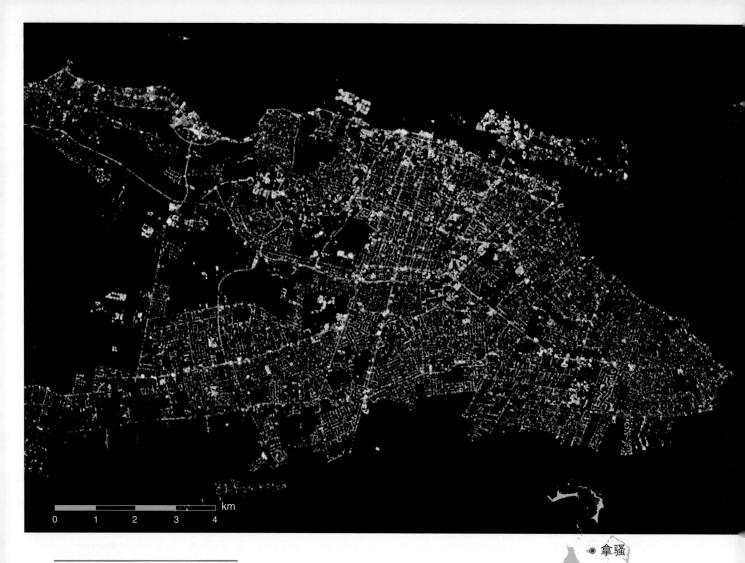

km
0   1   2   3   4

巴哈马 拿骚

Nassau, Bahamas

◉ 拿骚

巴哈马

拿骚是巴哈马的首都和第一大城市，城市面积约为 207 km²，位于新普罗维登斯岛上的拿骚是国家的商业、教育、法律、行政和媒体中心。林登平德林国际机场是巴哈马的主要机场，位于拿骚市中心以西约 16 km 处，每天都有航班飞往加拿大、加勒比海地区、英国和美国的主要城市。拿骚是众议院和各司法部门的所在地，也是国际金融中心之一。

Nassau is the capital and largest city of the Bahamas, with a city area of about 207 km². Located on the island of New Providence, it is the center of commerce, education, law, administration, and media of the country. Lynden Pindling International Airport, the major airport for the Bahamas, is located about 16 km west of the city center of Nassau, and has daily flights to major cities in Canada, the Caribbean, the United Kingdom and the United States. Nassau is the site of the House of Assembly and various judicial departments and is one of the international financial centers.

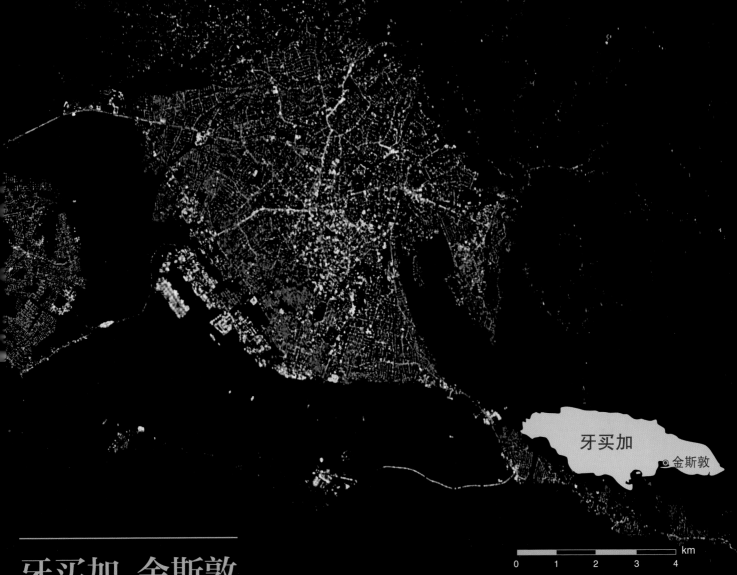

牙买加

金斯敦

牙买加 金斯敦

Kingston, Jamaica

　　金斯敦是牙买加的首都，位于牙买加的南岸，毗邻加勒比海，北靠蓝山山脉。金斯敦是牙买加最重要的港口城市之一，地理位置优越，便于与其他国家开展贸易和交流。金斯敦始建于 17 世纪末，19 世纪初正式建市，1872 年成为牙买加首都，其后迅速发展壮大。如今，金斯敦是加勒比地区的主要大都市之一，与皇家港和诺曼·曼利国际机场等重要交通枢纽相连，加强了与岛上其他地区的交流。金斯顿是牙买加的政治、经济和文化中心，每年吸引大量游客，促进了当地商业的发展。金斯敦人口约 59.1 万，市区面积约 25 km²。

Kingston is the capital of Jamaica, located on the southern coast of the island, adjacent to the Caribbean Sea, and bordered to the north by the Blue Mountains. It is one of Jamaica's most important port cities, benefiting from its strategic geographical location, facilitating trade and communication with other countries. Kingston was established in the late 17th century and officially chartered as a city in the early 19th century. In 1872, it became the capital of Jamaica, experiencing rapid growth and development in the years that followed. Today, Kingston is one of the major metropolitan areas in the Caribbean region, connected to important ports like Port Royal and the Norman Manley International Airport, enhancing international exchange with other regions on the island. It serves as the political, economic, and cultural hub of Jamaica, attracting a large number of tourists and fostering significant local commercial activities. The population of Kingston is about 0.591 million, and the urban area is about 25 km².

OCEANIA 大洋洲

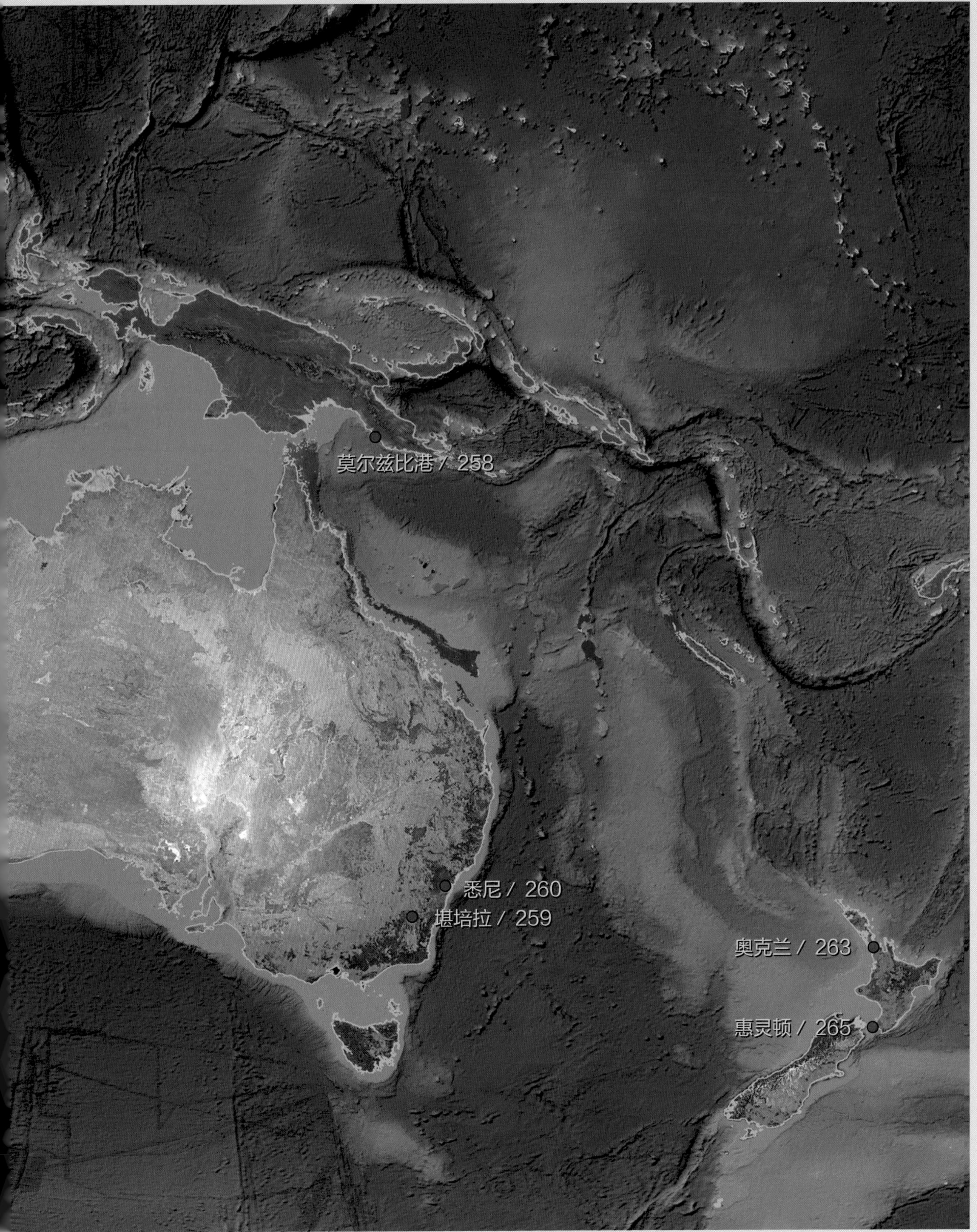

莫尔兹比港 / 258

悉尼 / 260

堪培拉 / 259

奥克兰 / 263

惠灵顿 / 265

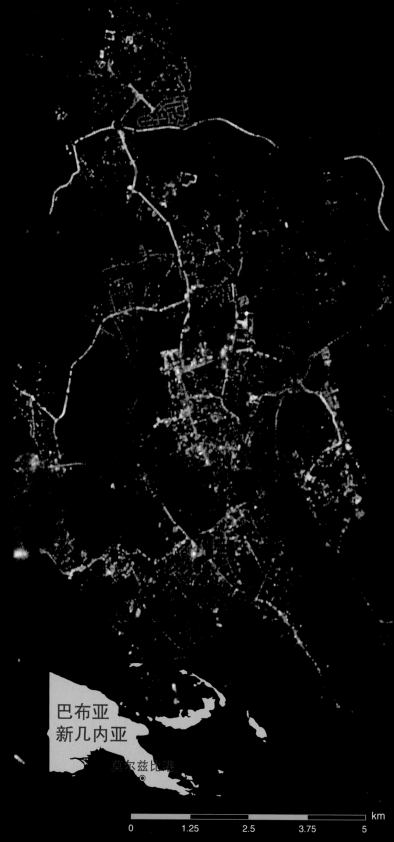

巴布亚新几内亚
莫尔兹比港
Port Moresby, Papua New Guinea

　　莫尔兹比港是巴布亚新几内亚的首都，是该国的门户。这座城市位于新几内亚岛的东南海岸，以其壮丽的港口和风景如画的地貌而闻名。这座城市拥有郁郁葱葱的绿地和美丽的海岸景色。作为国家主要的经济和政治中心，莫尔兹比港是政府机构的所在地和各种商业活动、国际赛事的举办地。游客可以探索当地市场，参观博物馆和文化遗址，体验巴布亚新几内亚独特的传统和文化。

　　Port Moresby is the capital city of Papua New Guinea and serves as a vibrant gateway to this diverse and culturally rich country. Situated on the southeastern coast of the island of New Guinea, the city is renowned for its magnificent harbor and picturesque landscapes. The city boasts beautiful natural scenery, with lush greenery and stunning coastal views. As the major economic and political center of the nation, Port Moresby hosts various government institutions, commercial activities, and international events. Visitors can explore local markets, museums, and cultural sites to experience the unique traditions and heritage of Papua New Guinea.

巴布亚
新几内亚

莫尔兹比港

km
0　　　1.25　　　2.5　　　3.75　　　5

澳大利亚 堪培拉
Canberra, Australia

澳大利亚

堪培拉
◎

0　　　1.25　　　2.5　　　3.75　　　5　km

　　堪培拉是澳大利亚的首都，也是其最大的内陆城市，位于澳大利亚最高山脉阿尔卑斯山脉的北端，在冬天可以看到白雪皑皑的山脉。堪培拉重视城市绿化，融入了大量的自然植被，被誉为大洋洲的"花园城市"，还被评为世界上最适合居住和旅游的城市之一。堪培拉是澳大利亚的政治中心，也是许多全国性社会和文化机构的所在地。

Canberra is the capital city and the largest inland city in Australia. The city is located at the northern tip of the Australian Alps, the country's highest mountain range, where snow-capped mountains can be seen in winter. Canberra places a strong emphasis on urban greenery, incorporating a significant amount of natural vegetation, earning it the title of the "Garden City" of Oceania. It has also been recognized as one of the world's most livable and tourist-friendly cities. Canberra is the political center of Australia and home to many national social and cultural institutions.

澳大利亚 悉尼
Sydney, Australia

　　悉尼是澳大利亚面积最大的城市，位于澳大利亚东南沿岸，是澳大利亚的经济、文化和旅游中心。悉尼也是全球重要的国际大都市，以其美丽的海滩、标志性的悉尼歌剧院和独特的悉尼海港大桥而闻名，拥有超过 70 个海港和海滩，包括全球最大的天然海港杰克逊港，以及著名的邦迪海滩。在繁华的中心商务区，摩天大楼与古老的砂岩建筑。在商务区南部，街道呈现整齐的网格状布局，与之形成对比的是古老的北部商务区街道较为复杂，反映了悉尼早期的街道发展特点。

　　Sydney, located on the southeastern coast of Australia, is the largest city in the country and serves as its economic, cultural, and tourist center. It is also a significant global metropolis, renowned for its beautiful beaches, iconic Sydney Opera House, and unique Sydney Harbour Bridge. With over 70 harbors and beaches, Sydney boasts the largest natural harbor in the world, Jackson Port, as well as the famous Bondi Beach. In the bustling central business district of Sydney, skyscrapers stand in contrast with ancient sandstone buildings. Towards the south of the business district, the streets form a neat grid layout, while in contrast, the older northern business district streets exhibit a more complex pattern, reflecting the early street development characteristics of Sydney.

澳大利亚

悉尼⊚

km

0   2.5   5   7.5   10

奥克兰

新

西

兰